Papers on Group Theory and Topology

Max Dehn

Papers on Group Theory and Topology

Translated and Introduced by
John Stillwell

With 151 Illustrations

Springer-Verlag
New York Berlin Heidelberg
London Paris Tokyo

John Stillwell
Department of Mathematics
Monash University, Clayton
Victoria 3168, Australia

AMS Classification: 01-A75, 55-03, 20 F 05, 20 F 32

Library of Congress Cataloging in Publication Data
Dehn, Max, 1878–1952.
  Papers on group theory and topology.
  Includes bibliographies.
  1. Combinatorial group theory—Collected works.
2. Topology—Collected works. I. Title.
QA171.D393A25   1987   512′.22   87-317

Printed and bounded by R.R. Donnelley & Sons, Harrisonburg, Virginia.
Printed in the United States of America

9 8 7 6 5 4 3 2 1

ISBN 0-387-96416-9 Springer-Verlag New York Berlin Heidelberg
ISBN 3-540-96416-9 Springer-Verlag Berlin Heidelberg New York

PREFACE

The work of Max Dehn (1878-1952) has been quietly influential in mathematics since the beginning of the 20th century. In 1900 he became the first to solve one of the famous Hilbert problems (the third, on the decomposition of polyhedra), in 1907 he collaborated with Heegaard to produce the first survey of topology, and in 1910 he began publishing his own investigations in topology and combinatorial group theory. His influence is apparent in the terms Dehn's algorithm, Dehn's lemma and Dehn surgery (and Dehnsche Gruppenbilder, generally known in English as Cayley diagrams), but direct access to his work has been difficult. No edition of his works has been produced, and some of his most important results were never published, at least not by him.

The present volume is a modest attempt to bring Dehn's work to a wider audience, particularly topologists and group theorists curious about the origins of their subject and interested in mining the sources for new ideas. It consists of English translations of eight works : five of Dehn's major papers in topology and combinatorial group theory, and three unpublished works which illuminate the published papers and contain some results not available elsewhere. In addition, I have written a short introduction to each work, summarising its contents and trying to establish its place among related works of Dehn and others, and I have added an appendix on the Dehn-Nielsen theorem (often known simply as Nielsen's theorem). The latter theorem was never published by Dehn, though Nielsen gives him credit for it, and Dehn's approach to the theorem became forgotten when Nielsen (1927) buried it in a 180-page paper. In fact, Dehn's approach can be made to work quite simply, as the appendix demonstrates.

In surface topology and the associated theory of fuchsian groups, Dehn's work fills a gap between the works of Poincaré and Nielsen. The three together are the main sources of contemporary work in this field, by Thurston in particular. With the recent publication of my translations of Poincaré's Papers on Fuchsian Functions (Springer-Verlag 1985) and Nielsen's Collected Mathematical Papers (Birkhäuser 1986), these sources are now available in what I hope is a convenient form.

The present volume would not have been possible without the generosity and encouragement of Wilhelm Magnus. He provided me with copies of important unpublished manuscripts of Dehn and put me in contact with Dehn's widow, Mrs. Toni Dehn, who very kindly gave permission for translations of these manuscripts to be published. Thanks are also due to B.G. Teubner and Co. for permission to publish the papers which originally appeared in Mathematische Annalen, to the Institut Mittag-Leffler for permission to publish the paper from Acta Mathematica, and to the Mathematisches Seminar of the University of Hamburg for permission to publish the Otto Schreier paper whose translation is included as an appendix to Paper 6.

The various papers in this book were typed by Anne-Marie Vandenberg and Joan Williams, at a time when publication was not foreseen. It is a tribute to the quality of their work that the book could be photographed directly from their typescripts. Wilhelm Magnus and Dave Johnson read large portions of the book and saved me from many errors. To all these friends and colleagues I offer my sincere thanks.

CONTENTS

Translator's Introduction 1   1

Paper 1 : Lectures on group theory   5

Translator's Introduction 2   47

Paper 2 : Lectures on surface topology   52

Translator's Introduction 3   86

Paper 3 : On the topology of three-dimensional space

    (Über die Topologie des dreidimensionalen Raumes.
    Math. Ann. 69, (1910), 137-168)   92

Translator's Introduction 4   127

Paper 4 : On infinite discontinuous groups

    (Über unendliche diskontinuierliche Gruppen.
    Math. Ann. 71 (1912), 116-144)   133

Translator's Introduction 5   179

Paper 5 : Transformation of curves on two-sided surfaces

    (Transformation der Kurven auf zweiseitigen Flächen.
    Math. Ann. 72 (1912), 413-421)   183

Translator's Introduction 6   200

Paper 6 : The two trefoil knots

    (Die beiden Kleeblattschlingen.
    Math. Ann. 75 (1914), 402-413)   203

Appendix to Paper 6 : On the groups $A^a B^b = 1$

    (Über die Gruppen $A^a B^b = 1$, by Otto Schreier,
    Abh. Math. Sem. Univ. Hamburg 3 (1924), 167-169)   224

Translator's Introduction 7   229

Paper 7 : On curve systems on two-sided surfaces,
    with application to the mapping problem.

    (Über Kurvensysteme auf zweiseitigen Flächen
    mit Anwendung auf das Abbildungsproblem.
    Vortrag (ergänzt) im math. Kolloquium,
    Breslau 11/2/1922)   234

Translator's Introduction 8   253

Paper 8 : The group of mapping classes

        (Die Gruppe der Abbildungsklassen.
        Acta Math. **69** (1938), 135-206)   256

Appendix : The Dehn-Nielsen theorem   363

TRANSLATOR'S INTRODUCTION

The following article is part of the first chapter of Dehn's lectures on group theory, probably written in 1909 or 1910. The date can be guessed from the next chapter of the notes, which includes an outline of some results from the famous paper, Dehn [1910], mentioning that details will appear in the Mathematische Annalen, which indeed they did in 1910.

The article is of interest mainly as Dehn's first discussion of the diagrams which he called "Gruppenbilder", and which are now known as Cayley diagrams. His first published work on these, Dehn [1910], concedes that Cayley had the idea first (Cayley [1878]), but it is clear from the present article that Dehn was not at all influenced by Cayley. He was, however, heavily influenced by the geometric represent-ations of groups which arise in function theory : the Poincaré [1882] definition of fuchsian groups by tessellations of the hyperbolic plane, and the more general theory of groups defined by tessellations in Dyck [1882].

Following Dyck and Poincaré, a tessellation which is mapped onto itself by all elements of a group G can be viewed as a "picture" of G when an arbitrary cell, R_1, is associated with the identity element 1 of G, and the cell R_g to which R_1 is sent by $g \in G$ is associated with g. The particular elements $g_1, \ldots, g_m$ which send R_1 to adjacent cells can be taken as generators of G, since R_1 can be sent to any R_g by repeatedly moving from one cell to an adjacent cell. Dehn's idea was simply to "dualise" this diagram : choose a point, called a <u>vertex</u>, V_g in each cell R_g to represent $g \in G$, draw edges from each vertex to the vertices in adjacent cells, and orient and label these edges with the corresponding generators.

An immediate advantage is that the <u>relations</u> of G are now highly visible, as closed edge paths in the diagram, and the relation corres-ponding to a given closed path can be immediately written down by reading the sequence of labels on its edges. In a Dyck style diagram, on the other hand, relations correspond to closed chains of cells, which are considerably harder to read.

Another advantage of Dehn's diagrams is that they can be defined without reference to the plane or other surfaces. The diagram D_G of G is just a labelled, directed, graph, with a vertex V_g for each $g \in G$ and, for each generator $g_i \in G$, an edge labelled g_i directed from V_g to V_{gg_i}. Dehn does not actually use this definition; perhaps because it is of no help in actually constructing diagrams. In practice, G is given by defining relations $r_1 = r_2 \ldots = r_n = 1$, and the whole problem is to decide what the vertex set of D_G is, i.e. to decide when two words in the generators are equal as a consequence of the relations.

There is no general way to do this, because of the unsolvability of the word problem (Novikov [1955]), but one can build "pieces" of D_G, each consisting of a vertex with closed paths $r_1, \ldots, r_m$ attached, and then hope that these pieces can be fitted together. Dehn's definition of the group diagram (section 1 below) reflects this state of affairs. At each vertex one has

    (1)   One incoming, and one outgoing, edge for each generator ;

    (2)   A closed path for each defining relation ("basic path").

A graph satisfying these conditions will in general represent only a quotient of G, since it may contain closed paths (relations) which are not put together (consequences) from the basic paths (defining relations). One therefore needs Dehn's third condition.:

    (3)   Each closed path in the diagram can be put together from basic paths.

Dehn does not define "put together" (zusammensetzen), but a suitable definition is evident when one considers the graphical counterpart of the process of inserting a defining relator in a word. Insertion of relator r between the subwords u,v of w = uv is achieved by

$w = uv \xrightarrow{\text{multiply by } v^{-1}rv} uv\,v^{-1}rv \xrightarrow{\text{cancel } vv^{-1}} urv,$ the graphical

counterpart of which is Fig. 1

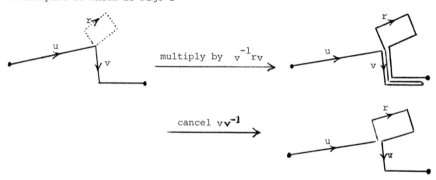

In other words "putting together" should mean forming products with paths of the form $v^{-1}rv$, where r is a basic path and v is arbitrary, and allowing trivial cancellation. One is then enabled to pull a path from one side to the other across any basic path.

Dehn's property 3 obviously holds for any planar or spherical group diagram, since the diagram then partitions the surface into cells whose boundaries are the basic paths, and any closed path can be reduced to the identity by pulling it across one cell at a time. Fortunately, many familiar and interesting groups are of this type, so Dehn is able to confine most of his exposition to this intuitively simple case. Most of the results he discusses were in fact already known, though of course he is putting them in a new and more unified setting.

The lectures are handwritten (not by Dehn) in a note book which for many years was in the possession of Dehn's student, Ruth Moufang. After her death it passed to Wilhelm Magnus, who very kindly provided me with a photocopy of the most interesting sections. The notebook is now with the rest of the Dehn Nachlass in the Humanities Library, Austin, Texas. The diagrams in the notebook are hand drawn and rather too small to reproduce well, so I have redrawn most of them, hoping that they will now be clearer, though still faithful to the originals. For extremely accurate and beautiful group diagrams one should consult the works of Fricke and Klein, or the commentary on them by Magnus [1974]. Some lesser known sources with splendid group diagrams are Burnside [1911], Threlfall [1932] and Coxeter [1939] (particular pp. 126-139) and [1979]. In a sense the _first_ non-trivial group diagram is in Schwarz [1872], though Schwarz does not make the group explicit. His diagram is also noteworthy because it does not hide the construction lines, something which all other authors, apart from Coxeter, have done.

REFERENCES

W. Burnside [1911] : Theory of Groups of Finite Order
2nd ed., Cambridge University Press

A. Cayley [1878] : On the theory of groups
Proc. Lond. Math. Soc. 9, 126-133

H.S.M. Coxeter [1939] : The abstract groups $G^{m,n,p}$
Trans. Amer. Math. Soc. 45, 73-150
[1979] : The non-euclidean symmetry of Escher's

picture 'Circle Limit III'
Leonardo 12, 19-25

M. Dehn [1910] :   Über die Topologie des dreidimensionalen
Raumes.
Math. Ann. 69, 137-168.

W. Dyck [1882] :   Gruppentheoretische Studien
Math. Ann. 20, 1-45

W. Magnus [1974] :   Noneuclidean Tesselations and Their Groups.
Academic Press.

P.S. Novikov [1955] :   On the algorithmic unsolvability of the word
problem in group theory (Russian)   Trudy Mat.
Inst. Steklov 44, 143 pp.

H.A. Schwarz [1872] :   Ueber diejenige Falle, in welchem die Gaussische
hypergeometrische Reihe eine algebraische Function
ihres vierten Elementes darstellt.
J. reine u. angew. Math. 75, 292 - 335.

W. Threlfall [1932] :   Gruppenbilder
Abh. sachs. Akad. Wiss. Math-phys. Kl. 41, 1-59.

LECTURES ON GROUP THEORY

GENERATION AND GEOMETRIC REPRESENTATION

1. Generation and geometric representation of $\Sigma_{3!}$

We begin our considerations with the generation and geometric representation of the group $\Sigma_{3!}$ of permutations of three things. We have
already shown that this group may be generated by a transposition s_1,
which exchanges the first and second terms, and a cyclic permutation
s_2 of all three terms, which replaces the first element by the second,
the second by the third, and then the third by the first. When we apply
the transposition s_1 twice, or the cycle s_2 three times, then in each
case we come back to the identity, hence

$$s_1^2 = 1 \quad \text{and} \quad s_2^3 = 1.$$

However, there must be another relation between these generating
elements s_1 and s_2, otherwise infinitely many elements could be composed
from them. This relation could say, e.g., that a certain power of the
product $s_1 s_2$ equals the identity. If the initial ordering of the three
things is 123, then the operation s_1 leads us to the ordering 213,
and if we apply the operation s_2 to this we come to the ordering 132;
this ordering becomes 312 after application of s_1, whence we return by
s_2 to the identity 123. Thus we have found that

$$(s_1 s_2)^2 = 1$$

Now we shall consider whether the group $\Sigma_{3!}$ is completely defined by the
generators s_1 and s_2 and the relations

$$s_1^2 = s_2^3 = (s_1 s_2)^2 = 1$$

between them. The answer to this question is given to us by the
geometric representation of a group, or the group diagram.

We take an arbitrary point [vertex], which will represent the identity,
and then two more vertices which will represent the orderings obtained from
the identity by the generators s_1 and s_2. The line segments [edges]
from the identity to these two points then represent the generators s_1 and
s_2 and we therefore label them by s_1 and s_2, or 1 and 2 for short,
and indicate their directions by arrows. When we again apply s_1 to the

vertex just obtained by the substitution s_1, we shall return to the identity since $s_1^2 = 1$. Thus the opposite direction on the edge 1 also corresponds to the element s_1, so that we can omit the arrow on this edge. However, this is not the case on the edge 2, because if we reapply the operation s_2 we do not come back to the identity, but to a new vertex. If we now apply the operation s_2 to this new vertex, then we do return to the identity, since $s_2^3 = 1$. Finally, we apply the operation s_2 to the vertex reached by the operation s_1 ; we then reach a new vertex, from which we must return to the identity by subsequent application of the substitutions s_1 and s_2, since $(s_1 s_2)^2 = 1$. The vertex at which we arrive by the

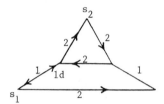

Fig. 1

substitution $s_1 s_2 s_1$ must coincide with the vertex for the substitution s_2^2 , because from both places we return to the identity by application of the operation s_2. When we represent all the different substitutions obtainable from s_1 and s_2 by composition, we come to a figure which is equivalent to the group generated by s_1 and s_2. This figure has the following properties :

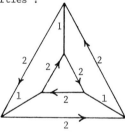

Fig. 2

1.) If we label each vertex with the operation by which it is reached from the identity, then one sees that the <u>neighbouring vertices of each vertex</u> <u>are precisely those obtainable by attachment of</u> $s_1, s_1^{-1}, s_2, s_2^{-1}$.

2.) <u>At an arbitrary vertex, regarded as the identity, the three rel-</u> <u>ations are satisfied</u>, since each vertex is in a quadrilateral, a triangle and 2-gon, i.e. a [double] line, and each relation between the generators corresponds to such a closed polygon.

3) <u>Each relation in the figure</u>, <u>i.e. each closed edge path</u>, can
be put together from the three basic relations, since each closed edge
path in the figure can be put together from basic polygons. E.g.

$$s_2^{-1} s_1 s_2 s_1 s_2^{-1} = s_2^{-1} (s_1 s_2 s_1 s_2) s_2^{-1} s_2^{-1} = s_2^{-1} s_2^{-1} s_2^{-1} = 1$$

<u>It follows from this property that our construction makes two different</u>
<u>elements of the group correspond to two different vertices, and conversely</u>,
so that the figure has exactly as many vertices as the group has elements.
Now, since our figure has six vertices, the group defined by s_1 and s_2
and the relations between them must be identical with $\Sigma_{3!}$. Because a
group which has additional relations must necessarily have fewer elements,
since elements which were different from the identity in the original
group now become equal to it.

The circumstances of one-to-one correspondence between group elements
and vertices, and the display of all possible relations between them,
entitle one to call such figures "<u>group diagrams</u>". We shall now see the
great advantages of such group diagrams in further examples.

2. <u>The group diagram of</u> $\Sigma_{4!}$. We now address ourselves to the problem
of constructing the group diagram for the group of permutations of four
things. Here we can choose the generating elements to be a transposition
s_1 and a four-termed cycle s_2, where the transposition always exchanges
the first and second elements and the cycle sends the first element to
the last place. Between s_1 and s_2 we then have the easily verified
relations

$$s_1^2 = s_2^4 = (s_1 s_2)^3 = 1.$$

The relation $s_1^2 = 1$ corresponds to a 2-gon in the group diagram, which
we represent by an edge without an arrow, the relation $s_2^4 = 1$ corresponds
to a quadrilateral, and the relation $(s_1 s_2)^3 = 1$ corresponds to a
hexagon. Thus, if the three relations are to be satisfied, we have to
draw the group diagram so that the three polygons in question meet at each
vertex. We then obtain a figure with 24 vertices, which assures us that
the group defined by s_1 and s_2 and the three relations is identical with
the symmetric group $\Sigma_{4!}$. One could also show this by mere calculation but
the process would be extraordinarily inconvenient. Our group diagram gives
us the result immediately ; it is to a certain extent a formula or

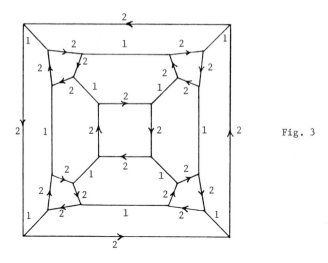

Fig. 3

computational scheme, and provides us with an example from the border
zone between formula and figure.

3. <u>Group diagrams for the alternating groups</u> A_{12} <u>and</u> A_{60}

Next we shall sketch the group diagram for the alternating group A_{12}.
Here we can proceed in different ways. On the one hand we can consider
two three termed cycles as generators, say

$$s_1 = (123) \quad \text{and } s_2 = (234)$$

for which we have the relations

$$s_1^3 = s_2^3 = (s_1 s_2)^2 = 1 \; ;$$

on the other hand, we can also take a double transposition with a three
termed cycle, e.g.

$$s_1 = (13)(24) \quad \text{and} \quad s_2 = (123),$$

for which the relations are then

$$s_1^2 = s_2^3 = (s_1 s_2)^3 = 1.$$

In the first case two triangles and a quadrilateral meet at each point,
in the second case there is a 2-gon, a triangle and a hexagon.

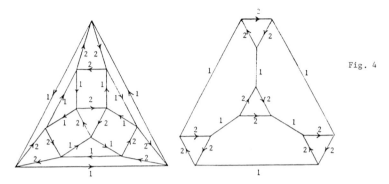

Fig. 4

Now we shall similarly draw the group diagram for the alternating group A_{60} of five symbols, which, as we know, consists of all even order permutations in $\Sigma_{5!}$. Since each odd cycle can be generated by an even number of transpositions, we choose the generators of our group to be a three-termed and five-termed cycle, say

$$s_1 = (123) \quad \text{and} \quad s_2 = (24153).$$

These two elements do in fact generate the alternating group A_{60}, since the group generated by them must be transitive and primitive; the first fact follows because it contains a five term cycle, the second because 5 is a prime number, because the permutations composed from s_1 and s_2 can only be of even order, and finally because a three-termed cycle appears in the group. Now the generators satisfy the relations

$$s_1^3 = s_2^5 = (s_1 s_2)^2 = 1$$

so that there is a triangle, a quadrilateral and a pentagon at each vertex.

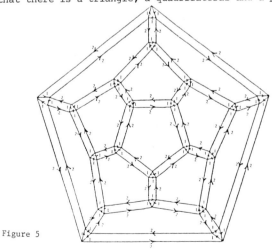

Figure 5

In order to give the figure the clearest possible form, we construct it from the dodecahedral net by replacing each vertex by a triangle whose sides correspond to the cycle s_1, and connect the vertices of the triangles in a suitable way by edges which correspond to the cycle s_2. The presentation of the alternating group A_{60}, which we have obtained in this way, was first given by Hamilton[*]

4. <u>The icosahedral and dodecahedral group</u>. The alternating group A_{60} coincides with the icosahedral and dodecahedral group since, as we shall now show, the latter have the same generators and relations as A_{60}. The three-termed cycle s_1 obviously corresponds to the rotation of the icosahedron about a face normal, i.e. about the line passing through the centres of two opposite faces, and the five-termed cycle corresponds to the rotation about a main diagonal, which connects two opposite vertices.

Now we have to establish the relations which hold between s_1 and s_2. If we consider one triangle of the icosahedron, say the triangle ABC, then s_1, the rotation about a face normal, sends A to B, B to C and C to A. If we then apply the operation s_2 i.e. rotate about the main diagonal BB' then the new point A remains fixed, while the new point B goes to the new point C and the point C goes to the point D [in the neighbour of the original triangle]. Thus the product s_1s_2 sends the original points A,B,C to the points B,A,D respectively. We see from the icosahedron figure that this change of positions can be brought about by an $180°$ rotation about the line connecting the midpoints of the edges AB and $A'B'$. If we now carry out a second such rotation, then we arrive back at the identity. Thus we have shown without special calculation that $(s_1s_2)^2=1$ and that a rotation about a line connecting midpoints of opposite edges can be obtained from the generators s_1 and s_2. Now, since generators s_1 and s_2 and the three relations.

$$s_1^3 = s_2^5 = (s_1s_2)^2 = 1$$

define a group with 60 elements, and since only 60 different rotations are possible for the icosahedron, we have in fact found the icosahedral group. We see from this that our generators and defining relations establish an isomorphism between the alternating group A_{60} and the icosahedral and dodecahedral group.

---

[*] Memorandum respecting a new system of roots of unity. Phil Mag. 12(1856) 446 (Translator's note).

5. Extension of the icosahedral group by reflections

We shall now derive, from our icosahedral group, a new group G_{120} of order 120, by combining the rotations of the icosahedron with its reflection in its midpoint. Such a reflection sends each vertex of the icosahedron to the diametrically opposite vertex. One sees without difficulty that such an exchange of vertices is not obtainable by rotation. Since one can apply this reflection to each of 60 positions obtainable by pure rotation of the icosahedron, the extension of the icosahedral group by reflection is a group with 120 elements. In order to be able to construct its group diagram we have to add a third generator s_3, representing the reflection in the midpoint, to the generators s_1 and s_2 of the icosahedral group. Since reflection in a point has period 2, i.e. two applications of the same reflection return each point to its original position, s_3 satisfies the equation

$$s_3^2 = 1.$$

The relations between s_3 and the two rotations s_1 and s_2 are also easy to determine. When we reflect the vertices of a triangle on the icosahedron, say ABC, then we obtain the diametrically opposite triangle and indeed A goes to A', B to B', and C goes to C'. If we then take the operation s_1, and thus rotate the triangle $A'B'C'$ about the normal through its midpoint, then A' goes to B', B' to C' and C' to A'. Finally, when we reflect the triangle $A'B'C'$, in its new position, the point A' goes to the old point B, B' goes to the old point C, and C' goes to the old point A. Thus the operation $s_3 s_1 s_3$ has sent the points A,B,C to B,C,A respectively, as also occurs with the rotation s_1 alone. Recalling that $s_3 = s_3^{-1}$, we then have the relation

$$s_3^{-1} s_1 s_3 = s_1$$

between s_1 and s_3. One can show similarly that s_2 and s_3 satisfy the same relation :

$$s_3^{-1} s_2 s_3 = s_2$$

Thus both the generators s_1 and s_3 are sent to themselves by transformation [conjugation] with s_3. We shall now show, with the help of the group diagram, that the group G_{120} is in fact completely defined by the operations s_1, s_2, s_3 and the six conditions

$$s_1^3 = s_2^5 = s_3^2 = 1, \quad (s_1 s_2)^2 = 1, \quad s_3^{-1} s_1 s_3 = s_1, \quad s_3^{-1} s_2 s_3 = s_2$$

We construct the group diagram from the group diagram of A_{60} by erecting perpendiculars of equal length, representing the generator s_3, at each vertex. We connect the endpoints of these perpendiculars so as to make another diagram of A_{60}. We then obtain a figure consisting of 120 vertices, whose top and bottom levels are both the figure for A_{60}. Thus the three relations $s_1^3 = s_2^5 = (s_1 s_2)^2 = 1$ are certainly satisfied at each vertex. It is also obvious that $s_3^2 = 1$, and the other two relations $s_3^{-1} s_1 s_3 = s_1$ and $s_3^{-1} s_2 s_3 = s_2$ are satisfied at each point as well, as one sees immediately by consideration of the small rectangles

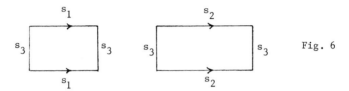

Fig. 6

formed by corresponding edges on the top and bottom levels and the connecting perpendiculars. Since our figure has the property that the number of edges issuing from each vertex is the same as the number of generators which can be applied, and the six relations between the generators are satisfied at each vertex, it is the diagram of the group defined by the generators s_1, s_2, s_3 and the six equations

$$s_1^3 = s_2^5 = s_3^2 = (s_1 s_2)^2 = 1, \quad s_3^{-1} s_1 s_3 = s_1, \quad s_3^{-1} s_2 s_3 = s_2.$$

It only has to be proved that all possible relations between s_1, s_2, s_3 in our group diagram can be put together from the six relations just mentioned. This holds not only for the top and bottom levels, where it is immediate since each is the diagram for A_{60}, but also for the whole figure, because each closed path in it bounds a simply connected piece of surface made up of pieces corresponding to relations. As in the well known Stokes' Theorem of integral calculus, a circuit round individual pieces is equivalent to a circuit of the boundary polygon of the whole surface, since inner edges are traversed twice, and in opposite senses, so that they cancel out. We have therefore proved completely that the six given relations between the generators s_1, s_2, s_3

completely define the group G_{120}, and that it is represented by our spatial group diagram.

6. **The group diagram of** $\Sigma_{5!}$. The extension of the icosahedral group by reflection is by no means identical with the symmetric group $\Sigma_{5!}$, even though they have the same number of elements. This will be shown by investigation of the latter group and its group diagram. As we know, the symmetric group $\Sigma_{5!}$ results from the alternating group A_{60} when we add a transposition s_3, which satisfies the equation

$$s_3^2 = 1.$$

Now we have only to establish the relations between s_3 and the other two generators s_1 and s_2, which are connected by the relation $(s_1 s_2)^2 = 1$, s_1 being the three-termed cycle (123) and s_2 the five-termed cycle (24153). If we choose s_3 to be the transposition (12), then simple calculation yields the easily verified relations

$$s_3^{-1} s_1 s_3 = s_1^2 = s_1^{-1} = (132) \quad \text{and} \quad s_3^{-1} s_2 s_3 = s_1 s_2^2 s_1^2 s_2^2 = (14253).$$

If the group G_{120} were isomorphic to the symmetric group on five symbols then the reflection would necessarily correspond to a transposition or a double transposition; because among the permutations of five symbols the only elements whose second powers equal the identity are the transpositions and double transpositions. On the other hand, the operation s_2 of G_{120} must correspond to an element of fifth order in the symmetric group, and in Σ_{120} this can only be a five-termed cycle. But now since we always have

$$t^{-1} c_5 t \neq c_5$$

in Σ_{120} when t is a transposition or double transposition and c_5 is a five-termed cycle (in fact, transformation by t leaves at least one term, say 1, unmoved, and then the term to which 1 is sent by c_5 is unmoved also, if $t^{-1} c_5 t$ is to equal c_5, and so on, so that t is the identity). It follows that G_{120} and Σ_{120} cannot be isomorphic groups.

We shall now use construction of the group diagram to show that the symmetric group Σ_{120} is completely defined by the generators

$$s_1 = (123), \quad s_2 = (24153), \quad s_3 = (12)$$

with the relations above. We set up the group diagram in a similar way
to that for G_{120} , by choosing the group diagram of the alternating
group A_{60} as top and bottom figure in a spatial structure whose
outline is that of a five-sided prism with parallel top and bottom
faces. The relations satisfied by s_1 and s_2 alone are then
satisfied at each of the 120 vertices of this figure. To satisfy the
remaining relations at each point, we connect each vertex of the bottom
figure with the points of the top figure associated with it by
$s_3^{-1} s_1 s_3 = s_1^{-1}$ and $s_3^{-1} s_2 s_3 = s_1 s_2^2 s_1^{-1} s_2^2$, and label the connecting
edges with s_3. It may then be shown that edges corresponding to
each of the five operations s_1, s_1^{-1} , $s_2, s_2^{-1}, s_3 = s_3^{-1}$ issue from
each vertex in the figure, also that the six given relations are
satisfied at each point and that each possible relation between the
generators in Σ_{120} can be expressed by these six basic relations.
This proves that our figure is in fact the group diagram for the symmetric group $\Sigma_{5!}$.

7. Generation of the group Σ_{120} by two operations.

In connection with the symmetric group Σ_{120}, we now take a somewhat
general view of the generation and geometric representation of groups.
Namely we pose the question whether the symmetric group Σ_{120} can be
generated by two generators. This is certainly possible, because we
shall show that a five-termed cycle s_2 and a transposition s_3 alone
suffice to bring each of the five symbols 1,2,3,4,5 to an arbitrary
position. We choose the five-termed cycle so that it brings the symbol
in the last place to the first place, first to second, etc., and the
transposition so that it exchanges first and second places. To
demonstrate our claim it suffices to prove that repeated application
of s_2 and s_3 suffices to bring an arbitrary symbol to a given
position, e.g. symbol 1 to the third place, without changing the order
of the other symbols.

Thus we have the original ordering :   12345
Order after application of s_3 :   21345
   "   "   "   " s_2^{-1} :   13452
   "   "   "   " s_3 :   31452
   "   "   "   " s_2 :   23145

This example discloses how we can generate the whole symmetric group Σ_{120} by application of s_2 and s_3 alone.

But now we still have to establish the relations for s_2 and s_3. First, we certainly have

$$s_2^5 = 1 \quad \text{and} \quad s_3^2 = 1$$

and the scheme :

Original ordering					12345
Order after application of				s_3	21345
"	"	"	"	s_2	52134
"	"	"	"	s_3	25134
"	"	"	"	s_2	42513
"	"	"	"	s_3	24513
"	"	"	"	s_2	32451
"	"	"	"	s_3	23451
"	"	"	"	s_2	12345

yields the relation

$$(s_3 s_2)^4 = 1$$

between s_2 and s_3. However, the three equations found so far are by no means sufficient to completely define the group Σ_{120}. This is certainly probable, in view of the fact that just now we needed six relations between the three generators, but nevertheless a rigorous proof is required. We do this by construction of the group diagram

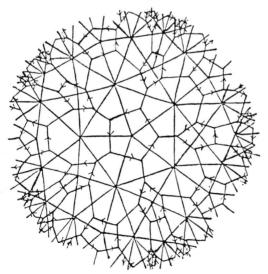

Fig. 7

When we draw the diagram of a group with generators s_2 and s_3 and relations $s_2^5 = s_3^2 = (s_3 s_2)^4 = 1$, a pentagon and two octagons must meet at each point, where the edges of the pentagons are denoted by s_2 and the edges common to two octagons are denoted by s_3. In our figure the s_2 edges are provided with arrows which give the direction of the operation s_2; the other edges need no indication of direction since $s_3 = s_3^{-1}$. (Here we are speaking only of the heavily drawn polygons; the lightly superimposed net of triangles will be explained later.) We now see from the constructed part of the group diagram that the group defined by s_2 and s_3 and the three relations certainly has more than 120 elements, because our figure already shows more than 120 vertices, from each of which issue three edges corresponding to the operations $s_2, s_2^{-1}, s_3 = s_3^{-1}$, and at each of which the three relations are satisfied. It may now be shown that the process of constructing two octagons and a pentagon at each vertex so as to satisfy the above relations, used for the polygons constructed so far, may be continued indefinitely. A rigorous proof of this can be given, either by elementary theorems of topology and induction, or else by methods of non-euclidean geometry. At any rate, it follows from this fact that the group defined by the generators s_2 and s_3 and the relations $s_2^5 = s_3^2 = (s_3 s_2)^4 = 1$ is of <u>infinite order</u>.

8. <u>Excursus into non-euclidean geometry.</u> Here we come for the first time to an infinite group. In the construction of group diagrams for infinite groups, a knowledge of non-euclidean geometry is indispensable. We shall therefore assemble the theorems of this discipline which are most important from our present point of view.

There are two different methods of representing the <u>non-euclidean plane</u>, one of which was developed by <u>Poincaré</u> in connection with the theory of automorphic functions, and the other of which was found directly by <u>Cayley</u> and <u>Klein</u>. Poincaré takes the non-euclidean plane to be the upper half of the ordinary euclidean plane bounded below by the so-called "Poincaré line". The "<u>non-euclidean lines</u>" in this half plane are the semicircles with midpoint on the Poincaré line, so the Poincaré line is cut orthogonally by all non-euclidean lines.

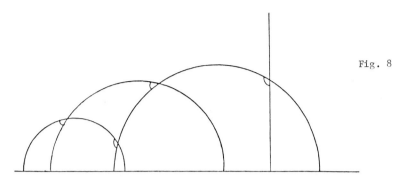

Fig. 8

Poincaré line

The justification of the term "line" for these semicircles is that such
a semicircle is uniquely determined by any two of its points, and two
such semicircles or non-euclidean lines have only one point of intersection [at most], since the lower half plane is omitted. The
"angle" between two intersecting non-euclidean lines is taken by
Poincaré to be the ordinary euclidean angle between the semicircles.

We shall now prove the following fundamental theorem of noneuclidean
geometry :

In a noneuclidean triangle, the angle sum is less than two right
angles.

Since inversion, or transformation by reciprocal radii, is known to
leave angles invariant, it suffices to prove our theorem for a right-
angled triangle whose short sides are segments of a perpendicular to the
Poincaré line and a semicircle bisected by this perpendicular. Thus we
have to prove our theorem for the triangle ABC , with right angle C ,
shown in the accompanying figure. It will then follow immediately for the
general right-angled triangle because we can always map the latter onto
one of our special right-angled triangles by inversion in one short side.
To prove the theorem we describe concentric circles around the foot of
the perpendicular. Then the angle between these circles and the hypotenuse
circle of our triangle increases until it takes its maximum value,
$\Theta = \theta_{max}$, at the point where the hypotenuse meets the perpendicular.
The most convenient way to see this is to replace the angle between the
circles by the angle between their radii. If we let ϕ denote the angle

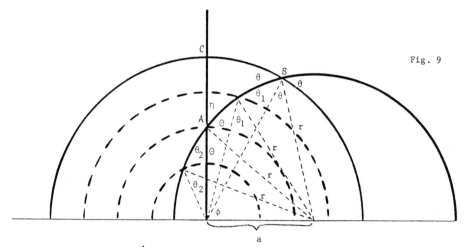

Fig. 9

between the Poincaré line and the variable radius vector to the
hypotenuse, let r be the radius of the hypotenuse circle and let
a be the distance from the centre of this circle to the foot of
the perpendicular, then the sine law of ordinary trigonometry gives

$$\sin \theta = \frac{a}{r} \sin \phi = c \sin \phi,$$

where c is a constant. Thus θ takes its maximum value for $\phi = 90°$,
and hence for the circle through the intersection of the hypotenuse and
the perpendicular. If we let η denote the angle in our right-angled
triangle at this point, then

$$\eta + \theta_{max} = \eta + \Theta = 90°$$

and consequently we must have the inequality

$$\eta + \theta < 90°$$

for our right angled triangle. This proves the assertion that the
angle sum of triangle ABC is less than two right angles.

Now, to prove this theorem for the angle sum of a general non-
euclidean triangle, we shall show that each triangle is decomposable
into two right-angled triangles. For this, we need only the following
lemma :

A triangle can have no more than one obtuse angle

For convenience' sake we again transform our triangle so the side
on which the hypothetical two obtuse angles lie is a segment of a

perpendicular ℓ to the Poincaré line. Now any circle whose lower
left angle with the perpendicular is obtuse must have its centre to
the left of ℓ, and hence it cannot cut the circle k in such a way
that the angles on ℓ of the resulting triangle are both obtuse.

Fig. 10

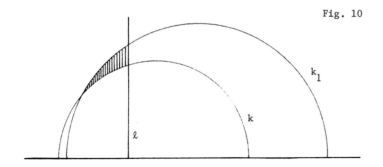

Now that we have shown that a triangle can have at most one obtuse
angle, we transform a given triangle so that the side opposite the
obtuse angle ACB goes onto a perpendicular to the Poincaré line,
and then take the foot of this perpendicular as the centre of a semicircle which goes through C

Fig. 11

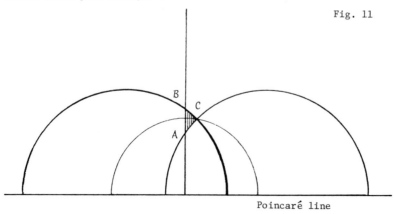

Poincaré line

Since this circle must always meet the base of our triangle, we have
divided the given triangle into two right-angled triangles, with
hypotenuses AC and BC. But now in each of these two right-angled
triangles the two angles at the hypotenuse sum to less than 90°,
hence the angle sum of the given non-euclidean triangle is less than two
right-angles, as was to be proved.

We can now go easily from the Poincare half plane by a conformal mapping to the <u>Klein-Cayley representation of the non-euclidean plane</u>. "Conformal" here means only that the Poincaré angles are sent to angles with the same value in the Cayley angle measure. We first map the Poincaré half plane onto a sphere whose south pole touches the half plane on the Poincaré line, by the inverse stereographic projection from the North Pole. This carries the Poincaré line into the vertical meridian circle coplanar with it, and the Poincaré half plane into the corresponding hemisphere. Since stereographic projection is conformal, the circles orthogonal to the Poincaré line became circles orthogonal to the main meridian。 These circles go to straight lines under parallel projection orthogonal to the plane of the main meridian, since our orthogonal circles are parallel to the direction of projection. The interior of the circle, onto which the hemisphere is projected, represents the non-euclidean plane according to Klein's method.

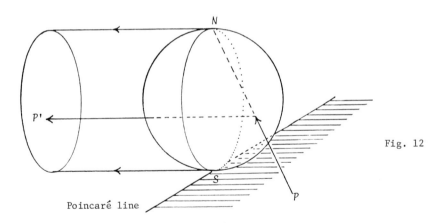

Fig. 12

Poincaré line

Each straight line in the latter disc corresponds to a circle orthogonal to the main meridian on the hemisphere and then to a circle orthogonal to the Poincaré line in the half plane. For lines through the midpoint of our absolute figure, which correspond to great circles on the sphere, the angles between them are preserved by the orthogonal parallel projection, though the angle in general is changed。 Each conformal circle-to-

circle mapping in the Poincaré half plane corresponds to a collineation
of the disc into itself. This enables us to carry over the measure of
angle between two non-euclidean lines to the Klein representation.
Namely, we can apply a collineation to send the two given lines to
lines through the midpoint, and then define the angle between the
original lines to be the angle between their transforms at the midpoint,
which equals the angle between the corresponding semicircles in the
Poincaré half plane.

It follows, first, that angle measures in both representations
of the non-euclidean plane agree. In particular, the Klein representation
also admits the important theorem that the angle sum of a non-euclidean
triangle is less than two right angles. But the collineations of the
disc into itself also yield a precise definition of equality of segments
in the non-euclidean plane. Namely, if we are given two segments AB
and $A'B'$ on the lines XY and $X'Y'$ then we can use two separate
collineations

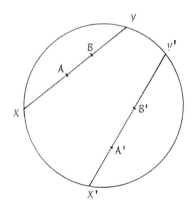

Fig. 13

to carry these lines, separately, to lines through the midpoint, then
bring the latter lines to the same position by rotation, and finally
bring the initial points of our segments into coincidence by a displacement
of one line along the other, when these three collineations bring the
segments AB and $A'B'$ into coincidence, we say they are equal. The
condition for this is equality of the two cross ratios $(ABXY)$ and
$(A'B'X'Y')$.

We now conclude our few details of non-euclidean geometry with a remark of special importance for the construction of group diagrams. A regular polygon in the non-euclidean plane can be continuously changed, without its ceasing to be regular, into one whose vertices lie on the boundary circle. One sees this immediately by consideration of polygons whose midpoints are at the centre of the circle, since they are simultaneously regular as both euclidean and non-euclidean figures.

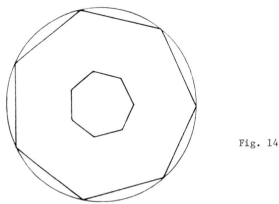

Fig. 14

But two lines meeting on the boundary circle enclose an angle of zero non-euclidean measure, since the corresponding semi-circles in the Poincaré half plane meet at the Poincaré line and therefore touch, so that the angle between them is zero. It follows from this, together with the theorem on the angle sum of a triangle, that in the non-euclidean plane the angle sum of a closed polygon with n sides can take any value between 0 and $(n - 2)\pi$.

9. <u>The group diagram of an infinite group.</u> Our last remark now provides us with a simple construction of the group diagram for the infinite group defined by generators s_2 and s_3 with the relations

$$s_2^5 = s_3^2 = (s_3 s_2)^4 = 1.$$

In this construction we apply the process of reflection in a line or circle, whereby a point A is replaced by its harmonic point A' with respect to G and S, where G is the pole of the line of reflection and S is the intersection of GA with this line. This reflection, which always maps the interior of the circle onto itself, is characterised by the fact that

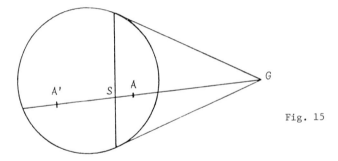

Fig. 15

the cross ratio $(G\ S\ A\ A') = -1$. Reflection obviously sends each closed polygon to a closed polygon, and each regular polygon to a regular polygon.

The group diagram of the above mentioned group must be set up so that a pentagon and two octagons meet at each vertex. Thus if we want to choose regular polygons in the ordinary euclidean plane, the rays corresponding to the operations s_2, s_2^{-1}, $s_3 = s_3^{-1}$ out of a given vertex will be such that the first two make an angle $3\pi/5$, while each makes an angle $3\pi/4$ with the third. But this is impossible in the ordinary plane, since $3\pi/5 + 2 \cdot 3\pi/4 > 2\pi$. In the non-euclidean plane, however, we can make the angle of regular polygon take any value between 0 and $\dfrac{(n-2)}{n}\ \pi$ by suitably choosing the side length, and hence we can also arrange for the sum of the angles of the pentagon and two octagons meeting at a vertex to be 2π. When this is done, the figure consisting of the regular pentagon and the two regular octagons immediately yields the whole group diagram, by reflecting the octagon in the edges corresponding to s_3.

Since this reflection preserves non-euclidean angles, five octagons always come to enclose a regular pentagon. As is easily proved, this process may be continued indefinitely without resulting in overlapping angles or incompatible labelling of edges. One sees immediately from the group diagram thus obtained that the group in question is of infinite order, as claimed. We also remark that the group diagram drawn by us was not obtained in the manner just described, however it gives an approximate conception of the actual figure produced by reflection.

10. Investigation of a certain class of groups

In connection with the infinite group just treated, we shall now investigate all groups defined by generators s_1 and s_2 and relations

$$s_1^\alpha = s_2^2 = (s_2 s_1)^\beta = 1$$

where α and β are any integers. We shall show that this group has infinitely many elements when the angle sum

$$\frac{\alpha-2}{\alpha} \pi + 2 \cdot \frac{2\beta-2}{2\beta} \pi \geqslant 2\pi.$$

Namely, in this case the group diagram has a non-euclidean regular α-gon and two non-euclidean regular 2β-gons at each vertex. Thus we construct the group diagram by choosing the side length of these regular polygons so that the sum of the three angles round a vertex equals 2π. By continued reflection of 2β-gon in those of its sides which correspond to the operation s_2, we obtain a group diagram of infinite extent, so that our groups are infinite in the cases mentioned.

On the other hand, when the above relation, which we can bring into the form,

$$2(\alpha + \beta) \leqslant \alpha\beta,$$

is not satisfied, then the elements s_1 and s_2 and relations $s_1^\alpha = s_2^2 = (s_2 s_1)^\beta = 1$ define only finite groups. If $\alpha = 2$ or $\beta = 2$ then $2(\alpha + \beta) > \alpha\beta$ always and the groups are finite : if $\alpha = 2$ the group diagram is a polygon with 2β edges, denoted alternatively by s_1 and s_2, and hence we have a group with 2β elements. The groups with $\beta = 2$ are isomorphic to these, since one has only to set $s_2 s_1 = \sigma_1$ to obtain the relations

$$\sigma_1^2 = s_2^2 = (s_2 \sigma_1)^\alpha = 1$$

For $\alpha = 3$ or $\beta = 3$ we have infinite groups when the inequality $6 \leqslant \beta$ resp. $6 \leqslant \alpha$ is satisfied, but we have to investigate the cases of β resp. α equal to $3,4,5$. When $\alpha = 4$ or $\beta = 4$ we get the inequality $4 \leqslant \beta$ resp. $4 \leqslant \alpha$, consequently we have only to investigate the case $\beta = 3$ resp. $\alpha = 3$ further. The same thing happens for $\alpha = 5$ or $\beta = 5$, since the inequality then takes the form $10 \leqslant 3\beta(\alpha)$. On the other hand, for $\alpha = 6$ or $\beta = 6$ our inequality is satisfied for all values $\alpha > 2$, $\beta > 2$, so that only infinite groups result.

Thus the only finite groups defined by generators s_1 and s_2 and relations

$$s_1^\alpha = s_2^2 = (s_2 s_1)^\beta$$

are those for which one of the integers α, β has the value 2, in which case finiteness is easily shown by construction of the group diagram, and the groups with the pairs of values

$$\alpha = 3, \ \beta = 3 \ ; \quad \alpha = 3, \ \beta = 4, \quad \text{or} \quad \alpha = 4, \ \beta = 3 \ ;$$

$$\alpha = 3, \ \beta = 5 \quad \text{or} \quad \alpha = 5, \ \beta = 3.$$

As we have already seen,

$$s_1^3 = s_2^2 = (s_2 s_1)^3 = 1$$

defines the alternating group A_{12}, which is isomorphic to the tetra-hedral group. The pairs of values $\alpha = 3, \ \beta = 4$ and $\alpha = 4, \ \beta = 3$ both give the same group, because if we introduce $s_2 s_1$ as a new generator s_1', then $s_2 s_1' = s_2 s_2 s_1$; and hence the relations $s_1^4 = s_2^2 = (s_2 s_1)^3 = 1$ may be replaced by $s_1'^3 = s_2^2 = (s_2 s_1')^4 = 1$. The group obtained here is the octahedral group, which indeed is iso-morphic to the symmetric group Σ_{24} on four symbols. The pairs of values $\alpha = 3, \ \beta = 5$ and $\alpha = 5, \ \beta = 3$, which again represent a single group because of the replaceability of $s_1^5 = s_2^2 = (s_2 s_1)^3 = 1$ by $s_1'^3 = s_2^2 = (s_2 s_1')^5 = 1$, lead to the icosahedral group, or altern-ating group A_{60} on five symbols.

11. **Group diagrams of the second kind.** By a kind of duality we can now move from the group diagrams previously discussed to "group diagrams of the second kind", which are very useful in certain circum-stances. Again we begin with the example of the infinite group defined by the relations $s_1^2 = s_2^5 = (s_1 s_2)^4 = 1$, whose group diagram has a pentagon and two octagons meeting at each vertex. In order to obtain the second kind of diagram from the diagram of this group prev-iously constructed, we mark the midpoint in each of the regular polygons from which the group diagram was constructed, and connect the midpoints of neighbouring polygons by straight lines. All these connecting lines, which bisect the polygon sides, form in our case the lightly drawn net of triangles which we have spread over the old group diagram [Fig. 7]. We obtain triangles here because three polygons meet at each vertex in the group diagram of the first kind ; in general, when n different

edges issue from each vertex of a group diagram of the first kind, then
the group diagram of the second kind consists of a net of n-gons.
Each vertex of the group diagram of the first kind is enclosed by such
an n-gon, and each polygon midpoint in the group diagram of the first
kind is the origin of as many edges in the diagram of the second kind
as the polygon has sides.

In our example, each vertex of the basic figure formed by the
non-euclidean regular pentagon and octagons is enclosed by a triangle.
Since eight such triangles must meet at the midpoint of each octagon, and
five must meet at the centre of each pentagon, the group diagram of the
second kind consists of a net of non-euclidean isosceles triangles in
which the angle at the pentagon midpoint is $\beta = \frac{2\pi}{5}$ and the angles at
the two octagon midpoints are $\alpha = \frac{2\pi}{8}$. The triangles are constructible
only in non-euclidean geometry, since the angle sum
$2\alpha + \beta = 2 . \frac{2\pi}{8} + \frac{2\pi}{5} = \frac{9}{10}\pi$ is less than π. Using such a non-
euclidean isosceles triangle , we can easily generate the whole net of
triangles, by repeated reflection in its sides. The resulting net of
triangles, which covers the whole non-euclidean plane, gives us the group
diagram of the second kind for our infinite group. The group is now
represented by associating each triangle with the element corresponding
to the vertex it encloses in the group diagram of the first kind. Then,
as one easily sees, the operation s_1 is represented by reflection in
the base, the operation s_2 by reflection in the right side, and s_2^{-1}
by reflection in the left side, of the triangle. One realises immediately, by consideration of the figure, that reflecting twice in the
base returns one to the initial position, but that five reflections
in the right side are required to do the same thing. The accompanying
fig. 16 shows the way in which individual triangles correspond to
group elements. One starts from an element denoted by 1 and writes
in each triangle the corresponding element. The operation s_2 has
been indicated by an arrow crossing the side where the corresponding
reflection occurs; since $s_1 = s_1^{-1}$ we omit arrows indicating reflection
in the base.

It may now be asked how one can use such a net of n-gons, uniquely
determined from the generators and defining relations, for the representation of finite [quotient] groups. To do this it is necessary to know
representations of the individual **elements** [coset representatives] of
the [quotient] group in terms of the generators. We then extend the

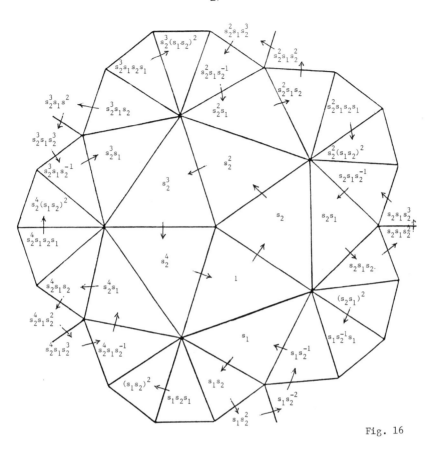

Fig. 16

polygon net just as far as is needed to associate each element with
exactly one polygon. Since the number of [quotient] group elements
is finite, the associated polygons form a finite surface, possibly
of higher connectivity, i.e. possessing holes, with a boundary
consisting of closed curves. We call this finite surface, which contains each element of the [quotient] group once and only once, the
<u>fundamental domain</u> of the [quotient] group*. Thus the fundamental
domain is a finite piece of an infinitely extended polygon net we shall
call an "<u>infinite basis</u>" for the finite [quotient] group. Now let
F be such a fundamental domain and let R be the boundary of this
finite surface F; for the sake of simplicity we assume that F is
made of triangles. Now consider a triangle a, one edge of which lies
on the boundary R of the surface F, and apply to it the operation s
corresponding to passage across this edge. We then arrive at a triangle
a_i on the surface [in the coset of as], one edge of which necessarily
lies on the boundary R. Because if this were not the case, application
of the operation s^{-1} would yield a triangle a' of F not coinciding
with the triangle a, whereas the element $a_i\, s^{-1} = ass^{-1}$ must be a.
The same element of the group would then be represented by two different
triangles in the fundamental domain, contrary to hypothesis. Consequently:
<u>in a fundamental domain the boundary edges correspond in pairs</u>.

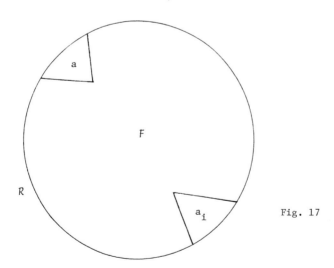

Fig. 17

<hr>

*Nowadays we would say it is the fundamental domain of the normal subgroup
by which the original group is divided in order to obtain the quotient
[Translators note].

In order to realise this correspondence geometrically, we shall
pull the fundamental domain together so that corresponding boundary
edges are united. We then obtain a closed surface in space, on which
each element of the group is represented by a polygon. For the six
elements making up the symmetric group $\Sigma_{3!}$ on three symbols, for
example, we have a fundamental domain of the form shown in the following two figures

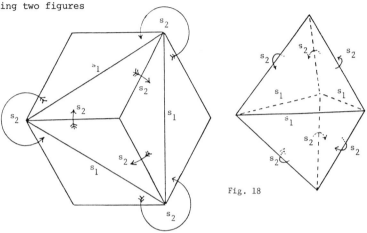

Fig. 18

The arrows denote the operation s_2, for which we must have $s_2^3 = 1$,
while the operation s_1, for which $s_1^2 = 1$, corresponds to passage
across the unmarked edges. The corresponding boundary edges are in
each case those cut by the same arrow. In order to reflect this correspondence, we have drawn together the fundamental domain so as to unite
corresponding boundary edges thus obtaining the group diagram of $\Sigma_{3!}$
in the form of a double pyramid with triangular base, or alternatively
the subdivision of the circumscribing sphere into six spherical
triangles. In general, of course, the surface in space will be much
more complicated than this simple example and usually of higher connectivity. It is obvious that in each case one can work backwards from
such a group diagram of the second kind to one of the first kind, by
the dualisation process we used in the first place.

13. <u>Sufficient relations for the generators.</u> Using the example of the
symmetric group Σ_{24} on four symbols, we shall now answer an important

question concerning the definition of groups by generators. We already know that each symmetric group can be generated by all the transpositions with a fixed term. Consequently, we can choose the three transpositions

$$s_1 = (12), \quad s_2 = (13), \quad s_3 = (14),$$

which satisfy the equations

$$s_1^2 = s_2^2 = s_3^2 = 1,$$

as generators for our group Σ_{24}. In order to construct the group diagram, we must set up a system of sufficient relations for these generators. We first find the relations

$$(s_1 s_2 s_3)^4 = 1 \quad \text{and} \quad (s_1 s_2)^3 = 1.$$

However, these relations are still not sufficient, since they define a group with infinitely many elements. This is easily shown by the group diagram of the second kind, which is very simple here. We shall not go into this, but we recommend proving this assertion as an exercise in the theory. What we are really interested in is the problem of finding a necessary and sufficient system of relations for the generators of a given finite group.

In order to be able to solve this problem more easily in general, we first consider the special problem of finding sufficient relations between the generators s_1, s_2, s_3 of the symmetric group Σ_{24}. We learned previously that the symmetric group Σ_{24} is also generated by a transposition t_1 and a four-termed cycle t_2, and that the relations

$$t_1^2 = t_2^4 = (t_1 t_2)^3 = 1$$

are necessary and sufficient to define it. Thus if we express t_1 and t_2 in terms of the new generators

$$s_1 = (12), \quad s_2 = (13), \quad s_3 = (14),$$

for which

$$s_1^2 = s_2^2 = s_3^2 = 1,$$

then we have

$$t_1 = s_1 \quad \text{and} \quad t_2 = s_1 s_2 s_3,$$

and we necessarily have the following relations for s_1, s_2, s_3 ;

$$t_2^4 = (s_1 s_2 s_3)^4 = 1 \quad \text{and} \quad (t_1 t_2)^3 = (s_2 s_3)^3 = 1.$$

However, as we have remarked above, these relations are certainly not sufficient. This is already clear on the grounds that s_2 and s_3 as yet appear together only as $s_2 s_3$, so that nothing has been said about them individually. In order to find sufficient relations, we shall represent s_1, s_2, s_3 in terms of t_1, t_2 ; we then have

$$s_1 = t_1, \quad s_2 = t_1^{-1} t_2^{-1} t_1 t_2 t_1, \quad s_3 = t_1^{-1} t_2^{-1} t_1 t_2^2,$$

as one easily checks by calculation. If we now replace t_1, t_2 by their expressions in s_1, s_2, s_3, then the latter equations become

$$s_1 = s_1, \quad s_2 = s_1 s_3 s_2 s_1 s_2 s_3 s_1, \quad s_3 = s_1 s_3 s_2 (s_1 s_2 s_3)^2.$$

These relations, in conjunction with the previous ones, now constitute a necessary and sufficient system for the generators s_1, s_2, s_3. We shall prove this assertion in a more general form.

Thus, let

$$s_1, s_2, s_3, \ldots, s_n$$

be generators of a group, which we know to be completely defined by generators

$$t_1, t_2, t_3, \ldots, t_m$$

and p relations

$$g_\pi(t_1, t_2, t_3, \ldots, t_m) = 1 \quad (\pi = 1, 2, 3, \ldots, p)$$

When we now express $t_1, t_2, \ldots, t_m$ in terms of $s_1, s_2, \ldots, s_n$:

$$t_\mu = \phi_\mu(s_1, s_2, s_3, \ldots, s_n) \quad (\mu = 1, 2, \ldots, m)$$

then the relations between $t_1, \ldots, t_m$ necessarily imply the relations

$$g_\pi(\phi_1, \phi_2, \ldots, \phi_m) = f_\pi(s_1, s_2, \ldots, s_n) = 1 \quad (\pi = 1, 2, \ldots, p)$$

between $s_1, s_2, \ldots, s_n$. If these relations are not yet sufficient, we represent $s_1, s_2, \ldots, s_n$ in terms of $t_1, t_2, \ldots, t_m$, obtaining

$$s_\nu = \psi_\nu(t_1, t_2, \ldots, t_m) \quad (\nu = 1, 2, \ldots, n).$$

If we replace the $t_1, t_2, \ldots t_m$ here by their expressions in $s_1, s_2, \ldots, s_n$, then we obtain the following additional relations for our new generators:

$$s_\nu = \psi_\nu(\phi_1, \phi_2, \ldots, \phi_m) = f_\nu^*(s_1, s_2, \ldots, s_n) \quad (\nu = 1, 2, \ldots n)$$

The $p+n$ relations

$$f_\pi(s_1, s_2, \ldots, s_n) = 1 \quad \text{and} \quad f_\nu^*(s_1, s_2, \ldots, s_n) = s_\nu$$

$(\nu = 1, 2, \ldots, n \ , \ \pi = 1, 2, \ldots, p)$

now give us a system of necessary and sufficient relations for the generators $s_1, s_2, \ldots s_n$.

To prove this we need only show that each relation possible
between $s_1, \ldots, s_n$ in the given group follows from our system. Suppose

$$R = s_1^{\alpha_1} s_2^{\beta_1} \ldots s_n^{\sigma_1} s_1^{\alpha_2} s_2^{\beta_2} \ldots s_n^{\sigma_2} \ldots s_1^{\alpha_r} s_2^{\beta_r} \ldots s_n^{\sigma_r} = 1$$

is any relation between $s_1, s_2, \ldots, s_n$ in our group. If we replace
$s_1, s_2, \ldots, s_n$ by their expressions $f_2^*(s_1, \ldots, s_n), \ldots, f_n^*(s_1, \ldots, s_n)$,
then we obtain

$$R = f_1^{*\alpha_1}(s_1, \ldots, s_n) f_2^{*\beta_1}(s_1, \ldots, s_n) \ldots f_n^{*\sigma_1}(s_1, s_2, \ldots, s_n)$$

$$\ldots f_n^{*\sigma_r}(s_1, s_2, \ldots, s_n)$$

$$= \psi_1^{\alpha_1}(\phi_1, \ldots, \phi_m) \psi_2^{\beta_1}(\phi_1, \ldots, \phi_m) \ldots \psi_n^{\sigma_1}(\phi_1, \ldots, \phi_m)$$

$$\ldots \psi_n^{\sigma_r}(\phi_1, \phi_2, \ldots, \phi_m)$$

$$= \psi_1^{\alpha_1}(t_1, \ldots, t_m) \psi_2^{\beta_1}(t_1, \ldots, t_m) \ldots \psi_n^{\sigma_1}(t_1, \ldots, t_m)$$

$$\ldots \psi_n^{\sigma_r}(t_1, t_2, \ldots, t_m)$$

But now, all relations in the group between $t_1, t_2, \ldots, t_m$ are consequences of the p relations $g_\pi(t_1, t_2, \ldots, t_m) = 1$.
We therefore have

$$\psi_1^{\alpha_1}(t_1, \ldots, t_m) \ldots \psi_n^{\sigma_r}(t_1, \ldots t_m) = \psi\{g_1(t_1, \ldots, t_m), \ldots, g_p(t_1, \ldots, t_m)\}$$

$$= \psi\{g_1(\phi_1, \ldots, \phi_m), \ldots, g_p(\phi_1, \ldots, \phi_m)\}$$

$$= \psi\{f_1(s_1, \ldots, s_n), \ldots, f_p(s_1, \ldots, s_n)\}$$

$$= 1$$

and our assertion is proved.

13. Groups with more than two generators

A few remarks remain to be made about the smallest number of generators of
a group. We have been able to generate all the finite groups considered
up to now by two elements, but it is easy to give examples of finite
groups   for which at least three generators are
required. These include, in particular, commutative or abelian groups.
As generators of such an abelian group we choose three elements s_1, s_2, s_3
and subject them to the six relations

$$s_1^2 = s_2^2 = s_3^2 = s_1 s_2 s_1^{-1} s_2^{-1} = s_1 s_3 s_1^{-1} s_3^{-1} = s_2 s_3 s_2^{-1} s_3^{-1} = 1$$

This group obviously cannot be generated by less than three elements, because if we choose two elements as generators, then either one generator is missing entirely, or else two of the elements s_1, s_2, s_3 appear only in combination, from which they cannot be determined individually.

Now our abelian group also has the special property, because of the exchangeability of the generators, of being decomposable. In general, a "decomposable group" is defined as the set of all products from two or more groups, where each element of one group commutes with each element of any other group . In our example the group decomposes into three groups, with the elements 1, s_1, resp. 1, s_2, resp. 1, s_3, each of which is of order 2 and generated by one element. We have the important theorem that : when a group G is decomposable into k groups $G_1, G_2, \ldots, G_k$, then G cannot have fewer than $n_1 + n_2 + \ldots + n_k$ generators*, and it has order $0_1 0_2 \ldots 0_k$, where n_i is the minimum number of generators of G_i and 0_i is its order, $i = 1, 2, \ldots, k$.

The group diagram of our abelian group of eighth order is a cube, whose eight vertices correspond to the eight elements, and the group diagram of the second kind is a regular octahedron, whose faces correspond to the individual elements.

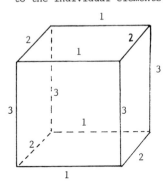
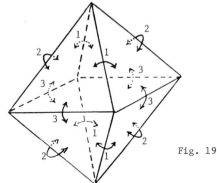

Fig. 19

We mention in passing that Holder has given an example, in the Mathematischen Annalen of 1893, of an indecomposable finite group which cannot be generated by two elements.

While there are finite groups which require more than two generators, they are nevertheless always subgroups of a group with two generators,

---

*This is false. E.g. $\mathbb{Z}_2 \times \mathbb{Z}_3 = \mathbb{Z}_6$ can be generated by one element. (Translator's note.)

namely the symmetric group. We can prove the theorem that :

> Each finite group G_n of order n is isomorphic to a subgroup of the symmetric group $\Sigma_{n!}$ of order $n!$

We denote the elements of G_n by

$$a_1, a_2, \ldots a_n$$

and introduce the notation

$$a_i a_k = a_{ik}$$

for the product of two of these elements. The element a_i can now be associated with the substitution

$$\begin{pmatrix} 1 & 2 & 3 & . & . & . & n \\ 1_i & 2_i & 3_i & . & . & . & n_i \end{pmatrix}$$

which replaces the indices of $a_1, \ldots a_n$ by those of $a_1 a_i, a_2 a_i, \ldots, a_n a_i$ respectively, where the latter sequence results from multiplication by a_i and hence also contains each element of G_n. In the same way, the element a_j is associated with the permutation

$$\begin{pmatrix} 1 & 2 & 3 & . & . & . & n \\ 1_j & 2_j & 3_j & . & . & . & n_j \end{pmatrix}$$

To prove our theorem, we now have only to show that

$$\begin{pmatrix} 1 & 2 & 3 & . & . & . & n \\ 1_i & 2_i & 3_i & . & . & . & n_i \end{pmatrix} \begin{pmatrix} 1 & 2 & 3 & . & . & . & n \\ 1_j & 2_j & 3_j & . & . & . & n_j \end{pmatrix} = \begin{pmatrix} 1 & 2 & 3 & . & . & . & n \\ 1_{(ij)} & 2_{(ij)} & 3_{(ij)} & . & . & . & n_{(ij)} \end{pmatrix}$$

or, since the left hand side equals

$$\begin{pmatrix} 1 & 2 & 3 & . & . & . & n \\ (1_i)_j & (2_i)_j & (3_i)_j & . & . & . & (n_i)_j \end{pmatrix}$$

that

$$(k_i)_j = k_{(ij)} .$$

But now $(k_i)_j$ is the index of $(a_k a_i) a_j$ and $k_{(ij)}$ is the index of $a_k (a_i a_j)$, and hence the associative law implies

$$(a_k a_i) a_j = a_k (a_i a_j) \quad \text{or} \quad (k_i)_j = k_{(ij)},$$

whence the correctness of our assertion, which has been extensively used by <u>Frobenius</u> and <u>Netto</u> in a series of important investigations.

A situation similar to that with finite groups, where the subgroup may need more generators than the group itself, also can be observed in infinite groups. As an example, we consider the infinite group with generators s_1 and s_2, and no relations between them. In the diagram of this group which consists of infinitely many vertices, from each of which issue four edges, corresponding to the operations $s_1, s_2, s_1^{-1}, s_2^{-1}$, there is no closed edge path. As subgroup of this infinite group we take the set of all elements resulting from s_1 by composition and transformation [conjugation] by arbitrary elements of the group itself. The elements of this subgroup, which is normal, cannot be generated by finitely many elements. We shall simply indicate the route to a simple proof of this assertion by means of an example. If we choose, e.g., $s_2^{-1} s_1 s_2$ and $s_2^{-2} s_1 s_2^2$ as generators, then they can generate only elements of the form

$$(s_2^{-1} s_1 s_2)^{\lambda_1} (s_2^{-2} s_1 s_2^2)^{\mu_1} \ldots (s_2^{-1} s_1 s_2)^{\lambda_\nu} (s_2^{-2} s_1 s_2^2)^{\mu_\nu}$$

$$= s_2^{-1} s_1^{\lambda_1} s_2^{-1} s_1^{\mu_1} s_2 \; s_1^{\lambda_2} s_2^{-1} s_1^{\mu_2} s_2 \ldots s_2 s_1^{\lambda_\nu} s_2^{-1} s_1^{\mu_\nu} s_2^2 \, ,$$

so that s_2 appears to no exponent greater than two. The element $s_2^{-\alpha} s_1 s_2^{\alpha}$, where $\alpha > 2$, can therefore not be generated.

14. <u>Hurwitz' investigations</u>. To conclude this chapter we shall now give, in a form somewhat modified to suit our purpose, a brief survey of the results published by <u>Hurwitz</u> in the Mathematische Annalen*. Hurwitz gives a quite specific rule for the construction of the fundamental domain <u>of a finite group</u> from the <u>infinite basis</u>**. If

$$s_1, s_2, \ldots, s_n$$

are the generators of the finite group, then Hurwitz introduces a new generator s_{n+1} by

$$s_1 s_2 s_3 \cdots s_n s_{n+1} = 1 \quad \text{or} \quad s_{n+1} = s_n^{-1} s_{n-1}^{-1} \cdots s_2^{-1} s_1^{-1} \, ,$$

so that the product of all generators is equal to 1. When the product $s_1 s_2 \cdots s_n$ is already 1, then of course it is unnecessary to add s_{n+1}. In any case, since our generators yield a finite group, equations of the form

$$s_1^{m_1} = s_2^{m_2} = \ldots = s_n^{m_n} = s_{n+1}^{m_{n+1}} = 1$$

---

*Vol.41(1893) 403-442 **This idea is actually due to Dyck, Math. Ann 20(1882) 1-45 (Translators note).

must hold, where $m_1, m_2, \ldots, m_n, m_{n+1}$ are finite integers, in addition to the relation

$$s_1 s_2 \cdots s_n s_{n+1} = 1$$

If the operations $s_1, s_2, \ldots, s_n, s_{n+1}$ have to satisfy only these equations then they generate an infinite group, whose group diagram of the first kind may be very easily constructed. At each vertex we need an m_1-gon with edges s_1, an m_2-gon with edges $s_2, \ldots$, an m_{n+1}-gon with edges s_{n+1} and between any two of these polygons in succession, an $(n+1)$-gon with edges $s_1, s_2, \ldots, s_{n+1}$, whose two edges at the given vertex are shared with the neighbouring polygons.

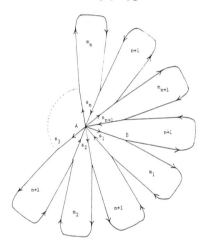

Fig.20

In our figure we have replaced these polygons by arcs which return to the initial point, and inside each arc we have inscribed the number of edges of the polygon. If these polygons are to be regular then of course they are only attainable in non-euclidean geometry, apart from the simplest cases.

If we now consider the point B, at which we arrive from A by means of the operation s_1, then an m_1-gon and an $(n+1)$-gon are already attached, next to each other, at this point. We must therefore fill the remaining angle around B with $(n+1)$-gons which alternate with an m_2-gon, and m_3-gon,..., an m_{n+1}-gon, i.e. a polygon net exactly the same as the one which fills the angle around A. The situation is similar at all other vertices, and it is clear how to continue the construction.

We can now proceed from this group diagram of the first kind to the group diagram of the second kind, by connecting midpoints of neighbouring polygons, whence the non-euclidean plane becomes covered with a net of (2n+2) - gons. If further relations are then given between the generators, and if the resulting group is finite, then we know how to mark off in this net a fundamental domain and join together its corresponding boundary sides. We then obtain the group diagram of the second kind for the finite group in the form of a closed surface, which in general will have higher genus. By dualisation we can convert this group diagram of the second kind into one of the first kind on the surface.

We now derive a simple relation between the order of the group and the genus of the surface resulting from the "Hurwitz basis", which carries the group diagram. Our starting point is the <u>Euler polyhedron formula</u>, already known from school :

<u>If α_0 is the number of vertices, α_1 the number of edges, and</u>
α_2 <u>the number of faces of a convex polygon,</u> <u>then</u>

$$\alpha_0 - \alpha_1 + \alpha_2 = 2$$

However, this formula holds only for polyhedra which can be continuously deformed into a sphere. For surfaces of higher connectivity, which have several holes, continuous distortion into a sphere is no longer possible. In this case we can obtain a simply connected surface by p cuts through the holes [and closing off the resulting perforations]. The cut open surface can be continuously deformed into a sphere, and hence the Euler polyhedron formula holds for it. But in cutting we have increased the number of faces by 2p, while changing the numbers of vertices and edges by the same amount, so that the number $\alpha_0 - \alpha_1$ remains the same. Hence for a surface with p holes we have

$$\alpha_0 - \alpha_1 + \alpha_2 = 2 - 2p,$$

assuming that the formula is correct for the sphere. In order to prove the latter, we show that the left hand side of this equation is independent of the subdivision of the surface into polygons. We can obviously convert one subdivision into another by either dividing edges in two [by new vertices] or by dividing faces in two by new edges. In the first case we we have one edge and one vertex more, in the second case one face and one edge more. The alternating sum

$\alpha_0 - \alpha_1 + \alpha_2$ therefore remains unaltered in both cases. Now for the simplest subdivision of the sphere, say by 2 caps ($\alpha_0 = \alpha_1$, $\alpha_2 = 2$) the formula $\alpha_0 - \alpha_1 + \alpha_2 = 2$ is immediate, and hence it also holds in general.

The important formula

$$\alpha_0 - \alpha_1 + \alpha_2 = 2 - 2p,$$

which relates the numbers of vertices, edges and faces of a subdivision to the genus of surface resulting from continuous deformation, now easily yields the Hurwitz theorem on the order of a group. If we consider the subdivision as the group diagram of the group generated by the operations $s_1, s_2, \ldots, s_{n+1}$, then the number of vertices, α_0, is the number of elements or the order N of the group, since $2(n + 1)$ edges issue from each vertex, the number of edges, α_1, is $N(n + 1)$. Finally, the number of faces is obviously equal to the sum

$$\frac{N}{m_1} + \frac{N}{m_2} + \ldots + \frac{N}{m_{n+1}} + \frac{N}{n+1}(n + 1) = N(1 + \sum_{i=1}^{n+1} \frac{1}{m_i}).$$

Our polyhedron formula therefore gives us the equation

$$N - N(n + 1) + N(1 + \sum_{i=1}^{n+1} \frac{1}{m_i}) = 2 - 2p$$

or

$$N\{ \sum_{i=1}^{n+1} (1 - \frac{1}{m_i}) - 2\} = 2p - 2.$$

This formula now permits all kinds of conclusions to be drawn concerning the order of the group, when we investigate individual surfaces in terms of their genus.

Suppose first that the group diagram is represented on a surface of genus $p > 1$, so that when $n > 3$ we have the equation

$$2p - 2 = N\{ \sum_{i=1}^{n+1} (1 - \frac{1}{m_i}) - 2\} \geqslant N \frac{n-3}{2} \geqslant \frac{N}{2},$$

since 2 is the smallest possible value of m_i. But it then follows that

$$p \geqslant 1 + \frac{N}{4}$$

and hence the genus of the surface increases with the order of the group. On the other hand, if we have $n = 3$ then not all the generators

s_1, s_2, s_3, s_4 can have order 2, since p is to be > 1. If we take one of them to be of order 3 and the others of order 2, then we obtain the equation

$$2p - 2 = N(\frac{2}{3} + \frac{3}{2} - 2) = \frac{N}{6}$$

or

$$p = 1 + \frac{N}{12},$$

and hence for all possible values of m_1, m_2, m_3, m_4 the genus of the surface must satisfy

$$p \geqslant 1 + \frac{N}{12} .$$

Finally, we still have to consider the case $n = 2$, for which we have the equation

$$N(3 - \frac{1}{m_1} - \frac{1}{m_2} - \frac{1}{m_3} - 2) = N(1 - \frac{1}{m_1} - \frac{1}{m_2} - \frac{1}{m_3}) = 2p - 2$$

since $p > 1$, no two of the three numbers m_1, m_2, m_3 can have the value 2, otherwise the left hand side would be negative. Also, they cannot all be 3, since p is to be greater than 1. For the same reason, when one of the three numbers, say m_1, is 2, we must still exclude the case $\frac{1}{m_2} + \frac{1}{m_3} \geqslant \frac{1}{2}$, which occurs for pairs satisfying $m_2 \leqslant 4$, $m_3 \leqslant 4$ or $m_2 \leqslant 3$, $m_3 \leqslant 6$. We have therefore established that when $p > 1$ and $n = 2$ the orders m_1, m_2, m_3 cannot be such that two of them have the value 2, or such that one has the value 2 and the others the values 3 and 3, or 3 and 4, or 4 and 4, or 3 and 5, or 3 and 6, or finally such that all three have the value 3. The smallest possible values of m_1, m_2, m_3 for $p > 1$ and $n = 2$ are therefore :

1) $m_1 = 2$, $m_2 = 3$, $m_3 = 7$; 2) $m_1 = 2$, $m_2 = 4$, $m_3 = 5$;

3) $m_1 = 3$, $m_2 = 3$, $m_3 = 4$; for which the corresponding values of p

are $1 + \frac{N}{84}$, $1 + \frac{N}{40}$, $1 + \frac{N}{24}$. Thus, if $p > 1$ <u>then it also satisfies</u> <u>the inequality</u>

$$p \geqslant 1 + \frac{N}{84},$$

which gives us the <u>Hurwitz result</u> on the order of the group and the genus of the group diagram.

Now that we have dealt with the surfaces of higher connectivity, we once again consider the case $p = 0$, in which the polygonal sub-division is on a convex surface deformable into a sphere. Here, our general formula yields the equation

$$N \left(\sum_{i=1}^{n+1} (1 - \frac{1}{m_i}) - 2 \right) = -2 \quad \text{or} \quad \sum_{i=1}^{n+1} (1 - \frac{1}{m_i}) = 2 - \frac{2}{N} \; .$$

Since $m_i = 2$ is the smallest possible value for the order of a generator, we certainly have

$$2 - \frac{2}{N} = \sum_{i=1}^{n+1} (1 - \frac{1}{m_i}) \geqslant \frac{n+1}{2} \quad \text{or} \quad n \leqslant 3 - \frac{4}{N} < 3$$

Thus the only possible values of n are

$$n = 1 \quad \text{and} \quad n = 2.$$

If $n = 1$ we have

$$2 - \frac{1}{m_1} - \frac{1}{m_2} = 2 - \frac{2}{N} \quad \text{or} \quad \frac{1}{m_1} + \frac{1}{m_2} = \frac{2}{N} \; ,$$

whence

$$N = \frac{2m_1 m_2}{m_1 + m_2} \; .$$

Now suppose m_2 is the greater of the two numbers m_1 and m_2, so certainly

$$\frac{2}{N} = \frac{1}{m_1} + \frac{1}{m_2} \leqslant \frac{2}{m_1} \quad \text{and} \quad \frac{2}{N} = \frac{1}{m_1} + \frac{1}{m_2} \geqslant \frac{2}{m_2}$$

whence

$$m_1 \leqslant N \leqslant m_2 \; .$$

On the other hand, since the order of the group itself must be greater than or equal to m_2, since it contains the m_2 elements $s_2, s_2^2, \ldots, s_2^{m_2}$, we necessarily have

$$N = m_2.$$

The equation $2 \dfrac{m_1 m_2}{m_1 + m_2} = N = m_2$ then yields

$$m_1 = m_2 \; .$$

so that in this case we have

$$m_1 = m_2 = N \; .$$

Consequently, our group is defined by

$$s_1^N = s_2^N = s_1 s_2 = 1 \; ,$$

where the relation s_2^N is in any case immediate from $s_2 = s_1^{-1}$, since $s_2^N = s_1^{-N} = 1$. The finite group obtained is therefore the "cyclic group" well known to us, or the "polygon group", whose group diagram can certainly be drawn on a sphere

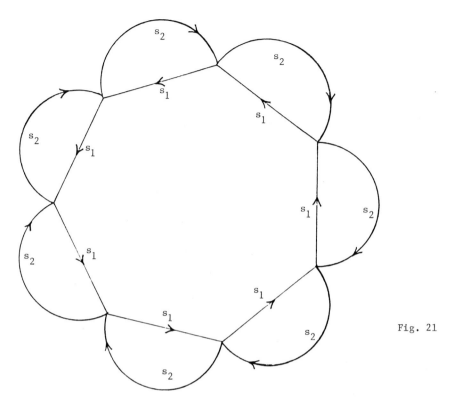

Fig. 21

Now we have to look at the case $p = 0$ and $n = 2$.

Here

$$3 - \frac{1}{m_1} - \frac{1}{m_2} - \frac{1}{m_3} = 2 - \frac{2}{N} \quad \text{or} \quad \frac{1}{m_1} + \frac{1}{m_2} + \frac{1}{m_3} = 1 + \frac{2}{N}$$

In order to investigate what cases are possible, we suppose that the ordering of m_1, m_2, m_3 is

$$m_1 \leqslant m_2 \leqslant m_3$$

We then have the following inequality for m_1 :

$$\frac{3}{m_1} \geqslant 1 + \frac{2}{N} \quad \text{or} \quad m_1 \leqslant \frac{3N}{N+2} < 3,$$

since N has to be a positive integer. We therefore have $m_1 = 2$ and hence

$$2 \leqslant m_2 \leqslant m_3 \quad \text{and} \quad \frac{1}{m_2} + \frac{1}{m_3} = \frac{1}{2} + \frac{2}{N} .$$

From the inequality

$$\frac{2}{m_2} \geqslant \frac{1}{2} + \frac{2}{N}$$

we find that

$$m_2 \leq \frac{4N}{N+4} < 4,$$

so that only the values 2 and 3 come into consideration for m_2. The equation

$$\frac{1}{m_1} + \frac{1}{m_2} + \frac{1}{m_3} = 1 + \frac{2}{N}$$

then enables m_3 to be calculated easily, and indeed $m_3 = \frac{N}{2}$ when

$$m_1 = m_2 = 2 \quad \text{and} \quad m_3 = \frac{6N}{N+12} = 6 - \frac{72}{N+12} = 6 - \frac{6}{\frac{N}{12} + 1} \quad \text{when}$$

$m_1 = 2$ and $m_2 = 3$. In the latter case m_3 is an integer only if

$\frac{N}{12} + 1$ equals $2, 3$ or 6, since N certainly has to be greater than zero. These three possibilities correspond to the values $m_3 = 3, 4, 5$, for which the N values are $N = 12, 24, 60$ respectively. Thus $p = 0$ the only possibilities for a group with three generators s_1, s_2, s_3 connected by the relation

$$s_1 s_2 s_3 = 1$$

are :

1) $\quad s_1^2 = s_2^2 = s_3^{m_3} = 1$

2) $\quad s_1^2 = s_2^3 = s_3^3 = 1 \; ; \; N = 12$

3) $\quad s_1^2 = s_2^3 = s_3^4 = 1 \; ; \; N = 24$

4) $\quad s_1^2 = s_2^3 = s_3^5 = 1 \; ; \; N = 60$

The first case is the dihedral, or double pyramid, group which is isomorphic to the group of rotations of a double pyramid whose base is a regular m_3 - gon. The operations s_1, s_2, s_3 respectively represent the two rotations which take one apex to the other and the rotation about the axis between the top and bottom apex. The remaining three cases lead us back to the frequently mentioned groups of the regular polyhedra, namely the tetrahedral, octahedral and icosahedral groups.

Thus we have completely settled the case $p = 0$; it has led us only to finite groups. The remaining case $p = 1$ gives only infinite groups as "bases", whose group diagrams are easily constructed, since they involve certain regular subdivisions of the euclidean plane.

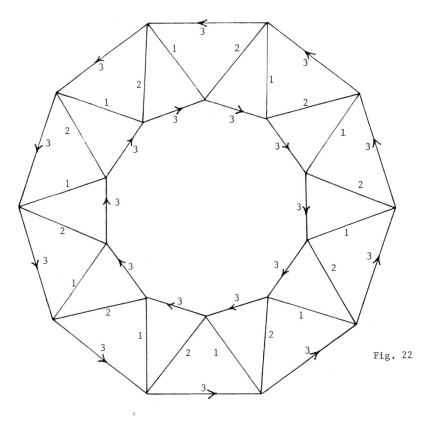

Fig. 22

In the case $p = 1$ our polyhedron formula becomes

$$\sum_{i=1}^{n+1} (1 - \frac{1}{m_i}) - 2 = 0 \quad \text{or} \quad \sum_{i=1}^{n+1} (1 - \frac{1}{m_i}) = 2,$$

the interesting aspect of which is that the order N of the group drops out entirely. Since 2 is the smallest possible value of the m_i, the number n satisfies the inequality

$$\frac{n+1}{2} \leqslant 2 \quad \text{or} \quad n \leqslant 3,$$

As one sees immediately, the equality sign here is only possible when

$$m_1 = m_2 = m_3 = m_4 = 2 \; ;$$

we shall look at the corresponding group again shortly. In the case $n = 2$ we have the equation

$$3 - \frac{1}{m_1} - \frac{1}{m_2} - \frac{1}{m_3} = 2 \quad \text{or} \quad \frac{1}{m_1} + \frac{1}{m_2} + \frac{1}{m_3} = 1$$

whose only solutions in integers m_1, m_2, m_3 are the triples

$$m_1 = 2, \; m_2 = 3, \; m_3 = 6 \; ; \quad m_1 = 2, \; m_2 = m_3 = 4; \quad m_1 = m_2 = m_3 = 3,$$

as is easily checked. A detailed treatment of the corresponding group given in the twelfth chapter of Burnside's Theory of Groups.

The minimal value of p for which the group diagram can be drawn on a closed surface of genus p is called the "genus of the group". Here there is a certain arbitrariness in the Hurwitz basis, since for certain groups one can find a representation on the sphere, whereas the Hurwitz method gives a genus $p > 0$. We shall now show a simple example of this.

We construct the group diagram for the infinite group with the four generators s_1, s_2, s_3, s_4 and the relations

$$s_1^2 = s_2^2 = s_3^2 = s_4^2 = s_1 s_2 s_3 s_4 = 1$$

The group diagram consists simply of rectangles which cover the whole plane. Since each vertex is the origin of as many edges as each cell has sides, our figure can be viewed simultaneously as a group diagram of the first and second kind. The first point of view takes the numbers on the edges as

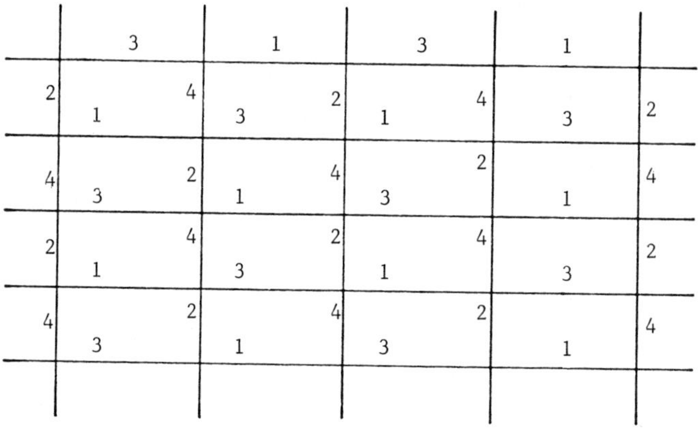

Fig. 23

operations by which one proceeds from one vertex to another, the second takes the operation to be the process of going from the cell on one side of the edge to the cell on the other. This diagram is the Hurwitz basis of the the finite abelian group G_8 given by the generators s_1, s_2, s_3 and the relations

$$s_1^2 = s_2^2 = s_3^2 = s_1 s_2 s_1^{-1} s_2^{-1} = s_1 s_3 s_1^{-1} s_3^{-1} = s_2 s_3 s_2^{-1} s_3^{-1} = 1.$$

Because if we introduce the generator s_4 by the equation

$$s_1 s_2 s_3 s_4 = 1,$$

then in fact

$$s_4^2 = 1$$

so that a fundamental domain for the abelian group G_8 may be found in our group diagram. This fundamental domain is made up of eight rectangles, and the corresponding edges on its boundary are those with the same labels.

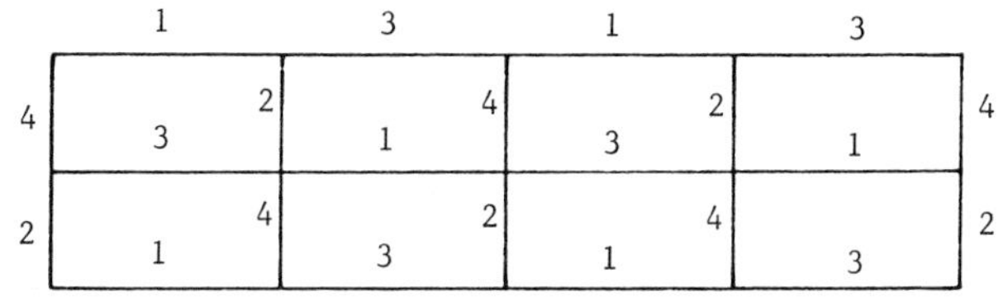

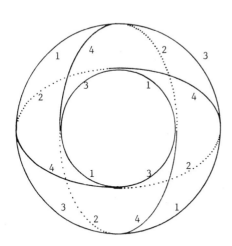

Fig. 24

By uniting the corresponding boundary edges one obtains the group diagram on a closed surface which has the form of a ring. While this

Fig. 25

gives the genus $p = 1$ for the group G_8, we have seen earlier that we can represent it in a much simpler way by a cube, and hence by a surface of genus $p = 0$.

We have now covered the most important aspects of the geometric representation of groups. Previously, these representations were used chiefly for intuitive support, but now they prove to be almost indispensable in the treatment of infinite groups, and consequently of increased significance.

The following chapter of Dehn's lecture notes deals mainly with
surface topology, presenting several of the results which were
eventually published in Dehn [1912a]. The account here is different
however, giving more details on non-orientable surfaces, and giving
different proofs of his main result, the solution of the conjugacy
problem (or "transformation problem" as Dehn calls it). These proofs
greatly illuminate the curious solution given in Dehn [1912a], which is a
combinatorial algorithm but justified by appeal to the hyperbolic metric,
revealing it as a transitional stage between the metric solutions and
the purely combinatorial solution given in Dehn [1912b].

The second proof given in the lectures is actually simpler than
the first, being an easy consequence of some ideas in Poincaré [1904],
a paper to which Dehn refers in [1912 a,b]. Poincaré takes the standard
universal covering $\widetilde{F}$ of a surface F of genus $p > 1$ by a net of
4p-gons in the hyperbolic plane, and lifts any non-contactible curve c
on F to a chain of curves ..., $\widetilde{c}^{(-1)}$, $\widetilde{c}^{(0)}$, $\widetilde{c}^{(1)}$,... on $\widetilde{F}$, where
the origin is the initial point of $\widetilde{c}^{(0)}$ and final point of $\widetilde{c}^{(i)}$ =
initial point of $\widetilde{c}^{(i+1)}$. The endpoints R,S of this chain on the
boundary circle $\partial\widetilde{F}$ are the fixed points of the motion, corresponding
to c, which maps $\widetilde{c}^{(i)}$ onto $\widetilde{c}^{(i+1)}$, and the hyperbolic line
connecting R,S is the axis A(c) of the motion (Fig. 1).
(Instead of the projective model used by Dehn in the previous paper,
I am now using the more familiar conformal model, in which "lines" are
circular arcs orthogonal to the boundary.)

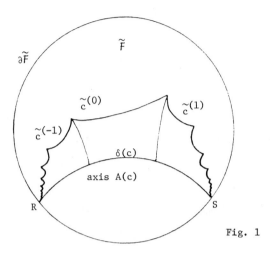

Fig. 1

Any lift in the chain, say $\tilde{c}{}^{(0)}$, can be deformed onto a displacement segment $\delta(c)$ of the axis while keeping the position of its ends equivalent, thus there is a corresponding transformation of c on F to the projection $\alpha(c)$ of $\delta(c)$. Moreover, any curve transformable into c is transformable into $\alpha(c)$, and conversely, hence a criterion for the transformability of one curve c into another, d, is that $\alpha(c) = \alpha(d)$. The curve $\alpha(c)$ is called the geodesic representative of the (free) homotopy class of c.

The important point is that $\alpha(c)$ is in a reasonable sense computable from c, so that one has an effective algorithm for solving the transformation problem. To compute $\alpha(c)$ one simply collects the segments of $\delta(c)$ from different polygons in the net and assembles their equivalents in a single polygon, e.g. as in Fig. 2.

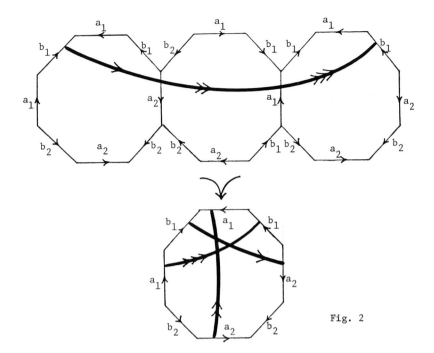

Fig. 2

The curve $\alpha(c)$ then corresponds to a sequence of directed arcs in a polygon, and is determined by the endpoints of these arcs on the polygon edges, which in turn are computable from the ends R,S of the axis A(c) and the position of the polygon net. Such calculations are not difficult in principle, nor do they seem to be complex in the sense in which one measures computational complexity nowadays, but it seems that the first and only time they were done was in the thesis of Dehn's student Gieseking in 1912. For more information on this work, and some explicit matrices for generating the coordinates of the net, see Magnus [1974].

Incidentally, the geodesic representative $\alpha(c)$ also gives a criterion for the transformability of c into a simple curve; namely, this is the case if and only if $\alpha(c)$ is simple. This was Poincaré's reason for looking at $\alpha(c)$.

Dehn's other solution of the transformation problem uses concepts from hyperbolic geometry, which also explain how to determine the ends R,S of the axis A(c). The motion with fixed points R,S which maps $\tilde{c}^{(i)}$ onto $\tilde{c}^{(i+1)}$ and A(c) onto itself is a displacement in the sense of hyperbolic geometry and therefore it maps each curve at a constant hyperbolic distance from A(c) (or distance curve of A(c)) into itself. Conversely, any sequence of points ... P_{-1}, P_0, P_1,... such that P_i is sent to P_{i+1} by the displacement must lie along a distance curve. In particular, this holds if

$$P_i = \text{final point of } \tilde{c}^{(i)}$$
$$= \text{initial point of } \tilde{c}^{(i+1)},$$

hence by taking three such points in succession we can construct the distance curve through them, and hence find R,S as its intersections with $\partial \tilde{F}$.

Dehn then decides whether two curves are transformable into each other by comparing distance curves rather than axes. This seems awkward and roundabout, particularly since he still needs the axes for his calculations, but it points the way to the next step, in Dehn [1912a], where axes and distance curves themselves are no longer needed, only certain properties they induce in the net of polygons, and finally to Dehn [1912b], where all metric notions are completely eliminated.

REFERENCES

M. Dehn [1912a]: Über unendliche diskontinuierliche Gruppen.
Math. Ann. 71, 116-144.

[1912b]: Transformation der Kurven auf zweiseitigen Flächen.
Math. Ann. 72, 413-421.

H. Poincaré [1904]:   Cinquième complément à l'analysis situs.

Rend. circ. mat. Palermo 18, 45-110.

W. Magnus [1974]:   Noneuclidean Tesselations and Their Groups

Academic Press.

LECTURES ON SURFACE TOPOLOGY

In the previous section we have discussed the generation of groups by given operations, and we have repeatedly come up against infinite groups thereby. In this chapter we shall now deal exclusively with such groups, and use them to make a series of important applications to analysis situs or topology.

§1. Closed surfaces

1. Preparatory theorems from topology. Before we treat the infinite groups themselves we must first get to know some preparatory theorems from topology. As we have already mentioned in the introduction, we can cut the <u>simple ring</u> [torus] along two curves a and b, one of which runs around the inner hole and the other across it, so that the torus surface can be spread out as a rectangle. The edges of this rectangle $ABCD$ correspond to the circuit cuts a and b, in fact each pair of opposite edges corresponds to one of the two cuts a, b. If we give each of the cuts a, b a definite direction, and denote the corresponding opposites by a^{-1}, b^{-1}, then a complete circuit of the rectangle $ABCD$ corresponds to traversal of the cuts of the torus surface in the order a, b, a^{-1}, b^{-1}. We must therefore denote the four edges AB, BC, CD, DA of the rectangle by a, b, a^{-1}, b^{-1} respectively.

Since our rectangle can obviously be regarded as a map of the torus surface, each curve Γ on the torus can be mapped onto the rectangle. At the places where Γ cuts the curves a and b its image will stop at a point on an edge of the rectangle, then continue at the corresponding ("homologous") point on the opposite edge. Consequently, a continuous curve on the torus appears on the rectangle divided into several pieces, whose endpoints lie on the edge in such a way that the final point of one piece and the initial point of the next are homologous points on opposite boundary edges. Conversely, when there is a system of curve segments given in the rectangle, pasting the rectangle together to form a torus brings homologous points on opposite edges together and the result is a continuous curve on the torus. This curve is then closed on the torus when its initial and final point coincide in the rectangle or, if they lie on the boundary, when they are at homologous points on opposite edges.

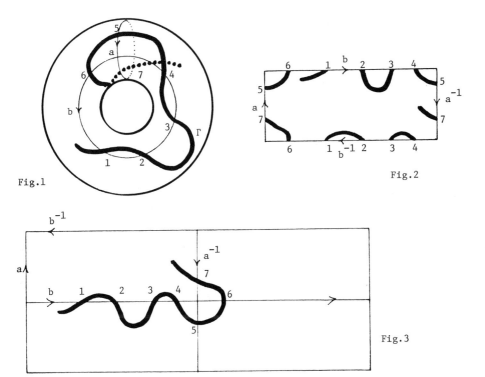

Fig.1

Fig.2

Fig.3

The mapping of curves on the torus becomes somewhat clearer when we no longer represent them by pieces, but instead return to continuous curves. To do this, a curve which meets an edge of the rectangle is continued, not at the homologous point on the opposite edge, but in a new rectangle on the other side of the edge being crossed. In this way we obtain a <u>net of infinitely many rectangles</u>, in which the individual pieces join together again into a connected curved path. This net of rectangles covers the plane simply and without gaps; four rectangles meet at each of its vertices, each rectangle has the four sides a, b, a^{-1}, b^{-1}, in the same order in each rectangle, and each vertex of the net is the origin of four segments a, b, a^{-1}, b^{-1}. Geometrically, the attachment of new rectangles is a multiple covering of the torus, the individual sheets of which meet along the curves a and b. If we are given a curve in the net of rectangles, then we can immediately find the corresponding curve on the torus by pasting together. The latter curve is then closed when the initial and final point of the image curve in the net correspond. This correspondence can indeed be set up arbitrarily in the construction of the net, but generally it is natural, for reasons of convenience, to let geometrically homologous points correspond.

We shall now similarly consider curves on the <u>double torus</u>. We cut this surface in the way already known to us along the four circuits a, b, c, d which have the positions shown in the figure. By spreading out the cut surface we then obtain an octagon as elementary surface piece, and we draw it as a regular octagon. With suitable choice of direction

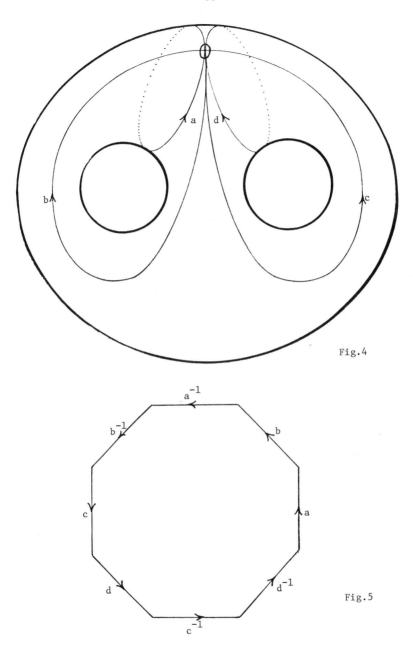

Fig.4

Fig.5

for the cuts a,b,c,d, the sides of the elementary octagon become the

sequence $a,b,a^{-1},b^{-1},c,d,c^{-1},d^{-1}$.

One finds this orientation when one traverses the cuts a,b,c,d

on the surface in such a way as to avoid crossing the point O. In

general, one sees that a closed surface of genus p can be spread out as

a plane elementary surface piece after cutting along 2p circuits

a_1, b_1, a_2, b_2,..., a_p, b_p, with a common point O and running alternately

through the holes of the surface and around them. This elementary

surface piece can be presented as a regular polygon of 4p sides, denoted

in order by a_1, b_1, a_1^{-1}, b_1^{-1},...,a_p, b_p, a_p^{-1}, b_p^{-1}. For this purpose

one chooses the orientation of the curves a_i and b_i, where a_i runs

through the ith hole and b_i runs around it, so that a circuit of the

polygon corresponds to a traversal of the cuts a_i, b_i on the surface

without crossing other cuts.

Each curve on the double torus which cuts none of the curves

a, b, c, d corresponds to a curve on the elementary surface piece. An

arbitrary curve on the double ring is represented by separate curve

pieces in the octagon, bounded by the points where the curve meets the

circuit cuts a,b,c,d. The boundary points are such that the final

point of one piece and the initial point of its successor are homologous

points on one of the edge pairs a and a^{-1}, b and b^{-1}, c and c^{-1},

d and d^{-1}. One can again replace these separate curve pieces by a

connected curve if one places a new octagon alongside each edge of the

original, with the same labelling and orientation. In the resulting net of octagons, eight polygons meet at each vertex. We must therefore set

things up so that octagon angle equals $\frac{2\pi}{8}$ = 45°. It follows that our

net of octagons can only be represented in the non-euclidean plane,

because there one can let the angle of a polygon take an arbitrarily

small value. The accompanying figure shows how the eight octagons lie

round a particular vertex in this non-euclidean net; the curved lines

are to be replaced by inscribed octagons.

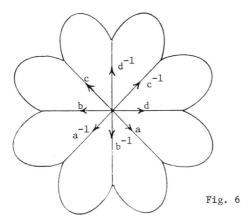

Fig. 6

This net of octagons admits a representation of curves on the double

torus similar to the representation of curves on the torus in the net

of rectangles. Each continuous curve on the surface maps to a continuous

curve on the net. Conversely, each curve on the net may be carried to

a curve on the double torus. The result is a closed curve on the

surface when the corresponding curve on the octagon net has homologous

initial and final point. Thus the net representation gives the means

of immediately deciding between closed and non-closed curves on the

surface. Naturally curves which are closed on both surface and net play

a special rôle. These closed curves have the property that they may
be continuously contracted to a point. Certainly, such a curve on the
net may be reduced to a point, but then each continuous movement on the
net corresponds to such a movement on the surface. Now, generalising
these considerations from the double torus to the surfaces of genus p
and the corresponding 4p-sided polygons which are their elementary
surface pieces, we obtain the following topological theorems:

1) Each curve on the surface which meets none of the circuit cuts
a_i or b_i corresponds to a curve within the elementary surface piece
which represents the surface. An arbitrary curve Γ divides into
segments bounded by the points where Γ meets the a_i, b_i; each such
segment corresponds to a segment on the elementary surface piece
connecting two boundary points, in such a way that the final point of one
segment and the initial point of its successor are homologous points on
segments a_i, a_i^{-1}, or b_i, b_i^{-1}.

2) If one constructs a net in the non-euclidean plane consisting
of 4p-sided regular polygons with angles $\frac{2\pi}{4p}$, and labels the edges of
these polygons, which cover the plane simply and without gaps, in the same
way and orientation as the elementary surface piece, as is always
possible, then each continuous curve Γ on the surface corresponds to a
continuous curve G on the net. If the curve Γ on the surface is closed
then the endpoints of the corresponding curve G are holomogous points
of the net, and conversely, when the endpoints of the net curve G
have homologous positions, the corresponding curve Γ on the surface is
closed. A closed curve on the net corresponds to a closed curve on the
surface which is contractible to a point.

The latter theorem may also be extended somewhat:

Each continuous transformation of a closed surface curve into another
may be assembled from separate operations in which at least one point
of the curve remains fixed.

Because, if Γ is a given closed curve on the surface we can
fix a point A on it and then continuously deform it into another
curve which has this point A in common with Γ and which also goes
through an arbitrary point B of the surface. In the same way, the new
curve can then be deformed into another while the point B remains fixed.
In our polygon net the curve Γ corresponds to a curve $G = A_1A_2$ whose
end points A_1 and A_2 are homologous net points which correspond to
the point A. The curve G is then deformed, first into another curve G'
which has initial and final points A_1 and A_2 in common with G
and which goes through the point B_1 [corresponding to the point B
(Translator's note)]. If we choose B as the fixed point of the curve,
then we must replace the piece A_1B_1 of G' by the homologous piece
A_2B_2 beginning at A_2, and we then have $G' = B_1B_2$. The curve G'
is then continuously deformed into the curve G^* which

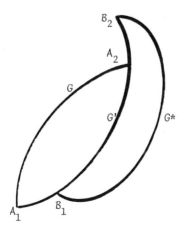

Fig. 7

has the endpoints B_1 and B_2 in common with G'. The net curves G and G^* then represent two curves Γ and Γ^* on the surface which are continuously deformable into each other.

2. Fundamental group of the surface of genus p

After these somewhat brief topological preparations, we now go to the fundamental groups of surfaces. The generators of the fundamental group of the surface of genus p will be called

$$a_1, \ b_1, \ a_2, \ b_2, \dots, \ a_p, \ b_p \ ,$$

for which a single relation holds:

$$a_1 b_1 a_1^{-1} b_1^{-1} a_2 b_2 a_2^{-1} b_2^{-1} \dots a_p b_p a_p^{-1} b_p^{-1} = 1$$

The group diagram of the group so defined is immediately seen to be equivalent to the net subdivision of the non-euclidean plane by 4p-sided regular polygons, or when $p = 1$, to the corresponding square net subdivision of the euclidean plane. The edges of an individual polygon are labelled by $a_1, b_1, a_1^{-1}, b_1^{-1}, \dots, a_p, b_p, a_p^{-1}, b_p^{-1}$, and in this order for all polygons. Issuing from each vertex of the polygon we have constructed are 4p edges, with the labels $a_1, b_1, a_1^{-1}, b_1^{-1}, \dots,$ $a_p, b_p, a_p^{-1}, b_p^{-1}$. Also, the single relation between the generators is satisfied at each vertex of the net, and each other relation resulting from the net is a consequence of this single relation, since each closed net curve may be put together from single polygons. Thus our polygon net is indeed the group diagram of the fundamental group of our surface.

With this group diagram we have already gained a great deal. First, we can immediately decide whether two given elements A and B of our group are identical or not. They are identical, i.e., $AB^{-1} = 1$, if and only if the points corresponding to them in the group diagram coincide. Thus the construction of the group diagram solves the most important problem, which we shall call the identity problem [now known as the word problem (translator's note)]; this is of great significance, since the equality of two elements of the fundamental group, given as products of generators, cannot be determined from the relation between a_i, b_i without extra work. Now, since our group diagram coincides with the mapping of our surface as a regular polygon net in the non-euclidean plane, we can also regard the vertices corresponding to elements of the group as the endpoints of net curves. If now the elements A and B are identical, or if $AB^{-1} = 1$, then the edge paths leading to them from the identity vertex form a closed curve in the net. Thus the identity problem is equivalent to the question of "null reduction", i.e., to the question of whether a closed curve on the surface may be continuously contracted to a point.

It is significantly more difficult to solve the transformation problem; its solution answers the question of whether two given elements A and B of the group can be transformed into each other by a third element C of the group, i.e. $A = C^{-1}BC$ [hence the current term, conjugacy problem (translator's note)]. This question proves to be equivalent to the deformability of closed curves Γ and Γ' on

the surface into each other. In order to see this, we first deform
the curves Γ and Γ' into curves which go through a point O
of the surface, and then replace them, by an inessential displacement,
by curves which follow the 2p circuit cuts. The possibility of such
continuous transformation of all closed curves on the surface follows
immediately from our last theorem in the topological preparations.
Now when we choose the point O as initial point, our new curves
correspond to certain edge paths in the polygon net , with their endpoints
at vertices. Now if $G = A_1A_2$ is such an edge path, with endpoints
A_1 and A_2, then when we choose another vertex B_1 as the initial
point of G we must attach to the point A_2 a piece A_2B_2 homologous
to A_1B_1, and we then obtain the edge path $B_1B_2 = H^{-1}GH$, where
$H = A_1B_1$. Thus by choosing a new initial point we have arrived at a
transformed edge path, which we now have to reattach to the initial
point A_1. We then obtain the two edge paths $G = A_1A_2$ and $K = A_1B_2'$,
where G is transformed into K as a result of the equation

$$K = H^{-1}GH.$$

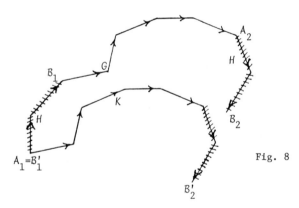

Fig. 8

Now we can regard the vertex A_1 as the identity element in the group diagram of our fundamental group, then the point A_2 corresponds to an element A and the point B_1 to an element C, and finally the point B_2' corresponds to an element B, so we have

$$B = C^{-1}AC.$$

Now since the curve K results from G by continuous transformation, we have shown that the transformation problem for the elements of our group corresponds to the question of when two closed curves on the surface are deformable into each other. Here we again notice the close connection between the fundamental group and the surface, to each element of the group corresponds a curve on the surface, so that we can almost represent the group by the surface itself.

3. Solution of the transformation problem. While the identity problem is immediately solved by the construction of the group diagram, a complete solution of the transformation problem requires a deeper investigation of the group diagram. The simplest diagram of a fundamental group that we have seen is the square net in the euclidean plane. Here we have only two generating operations, a and b, with the relation

$$aba^{-1}b^{-1} = 1.$$

We notice immediately from the validity of the commutative law ab = ba that our group is abelian and consequently representable symbolically by

$$T_1 = \{a^n b^m\},$$

in which m and n are positive or negative integers.

Now let

$$A = a^{n_1} b^{m_1} \quad \text{and} \quad B = a^{n_2} b^{m_2}$$

be two elements which are transformable into each other by the element

$$C = a^{\nu} b^{\mu},$$

then we have

$$a^{n_1} b^{m_1} = a^{-\nu} b^{-\mu} a^{n_2} b^{m_2} a^{\nu} b^{\mu}$$

or, because of the validity of the commutative law,

$$a^{n_1} b^{m_1} = a^{n_2} b^{m_2}.$$

But then it follows by consideration of the diagram that

$$n_1 = n_2 \quad \text{and} \quad m_1 = m_2.$$

Thus, <u>in the fundamental group of the torus,</u> two elements are transformable into each other if and only if they are identical. Thus the transformation problem here is trivial and without beauty.

Our problem becomes much more interesting in the general case. Here the relations between the points A_2 and B_2' (Fig. 8) have to be investigated. We shall now denote the polygonal path $A_1 A_2$ by A, and the polygonal path $A_1 B_2'$ by B, and we shall represent both by straight lines for the sake of simplicity. If we now move the net onto itself in such a way that the point A_1 goes to A_2 and each edge goes to a like-labelled edge, then the point A_2 goes to A_3, A_3 to A_4, A_4 to A_5, ..., A_0 to A_1, A_{-1} to A_0, ... and all these points are at distance A from each other; they constitute a chain of equal sides, which we shall call the A-chain for short. The angles between successive sides in this A-chain are equal, since the motion of the net carries the point A_i to A_{i+1} and hence brings the triangle $A_{i-1} A_i A_{i+1}$ into coincidence with $A_i A_{i+1} A_{i+2}$. But it follows from this that these points A_i all lie on a non-euclidean circle, whose midpoint M can be found, say, by bisecting the angles at two points A_i. This midpoint

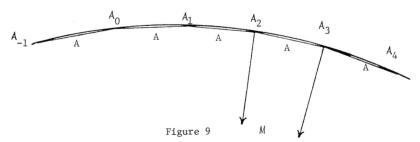

Figure 9

proves to be ideal, i.e. it lies outside the non-euclidean plane. If it were real, then the motion of the net, which is obviously a rotation about M, would bring a netpoint sufficiently close to M onto an arbitrarily close point. However, this is impossible, since a non-identity motion of the net sends each point inside a polygon to a homologous point in another polygon, and two such homologous points are always at a finite distance from each other [greater, e.g., than the length of a net edge (translator's note)]. For curves on the surface, this means that there are no infinitely

small closed curves which cannot be contracted to a point. Since
the circle in which the A-chain is inscribed has an imaginary or
ideal midpoint, it must meet the absolute, i.e. the boundary circle
of the non-euclidean plane, in two distinct points R and S.
This implies, by the way, the important property that no power of
A can equal 1, without A itself being equal to 1; of course such
a property is impossible in a finite group. The line connecting
these two infinitely distant points will be called the "<u>axis of the
motion</u>".

This axis is nothing but the polar of the point M with
respect to the absolute. For all points A of an ellipse which
touches the absolute at R and S, the cross-ratio of the four
points M,P,A,Q is constant, as we know from analytic geometry.
We are therefore justified in
regarding such ellipses as
non-euclidean circles with
ideal midpoint M. All
such non-euclidean circles
with the same ideal mid-
point have the same real
axis in common.

If Γ is the
geometric connection of
the point A_1 to the
point B_1 which corresponds to the element C, then the net motion

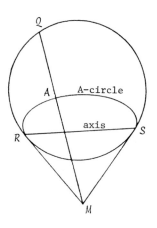

Figure 10

represented by the rotation about the ideal midpoint M sends the point B_1 to B_2, B_2 to B_3, ..., B_0 to B_1, B_{-1} to B_0,... . The points B_i thereby have distance Γ from the points A_i and the angles $B_i A_i A_{i+1}$ are all equal. The points ..., $B_{-1}, B_0, B_1, B_2, ...$ which constitute a $\Gamma^{-1} A \Gamma$-chain, therefore also lie on a non-euclidean circle which is concentric with the A-circle and has the same axis RS . Moreover, the ideal angle of rotation about M is the same for pairs of successive points on the A-chain and $\Gamma^{-1} A \Gamma$-chain respectively.

We can now briefly summarise our various results in the following theorems:

1. <u>In a motion of the net which brings like-labelled edges into coincidence and carries the vertex</u> A_1 <u>to</u> A_2 , <u>all vertices of the resulting A-chain lie on a circle, with an ideal midpoint, which touches the absolute at two distinct points, whose connecting line is called the axis of the motion.</u>

2. <u>Under this motion, all points of a</u> $\Gamma^{-1} A \Gamma$-<u>chain lie on a circle which has the same midpoint, axis and angle of rotation as the A-chain.</u>

With the help of these theorems the transformation problem can now be solved without too much difficulty. Let A and B be the elements under investigation, whose straight line connections with the identity element, which is represented by the point $A_1 = B_1'$, will be denoted by $A = A_1 A_2$ and $B = B_1' B_2'$. First we construct the A-chain and B-chain and their circumscribing circles, which

both have ideal midpoints and angles of rotation, but real axes.
Then we describe the non-euclidean circle with the same midpoint
as the A-chain and the same radius as the B-chain, so that the
axis of the [new] B-circle coincides with that of the A-circle.
However, since these circles have imaginary midpoints, we must
construct the new circle by determining all the points which have
the same distances d_b from the A-axis as the points of the [old]
B-circle have from the B-axis. [Actually, d_b will be constant,
as a non-euclidean distance (Translator's note).]

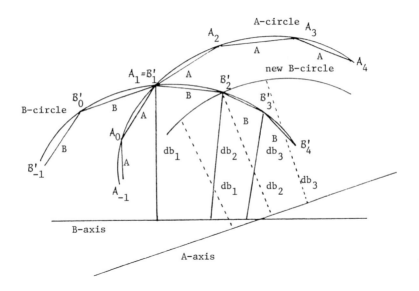

Figure 11

Now if A is transformable into B, the new B-circle must
satisfy two requirements vis à vis the axis of the A-circle: in
the first place, a circle K of radius $\dfrac{|B|}{3}$ around one of its
vertices must contain at least one vertex Y such that the edge

path B issuing from V ends at a vertex H on the new B-circle. Secondly, when this necessary condition is satisfied, the ideal angle of rotation of the motion $V \to H$ must agree in sense and magnitude with the ideal angle of rotation of the motion $A_1 \to A_2$.

In this way the identity and transformation problem can be solved by geometric operations. The analytic execution of this process is still called for, and could for that reason be the subject of a longer work.

4. Second method for the solution of the transformation problem.
We shall now give a second general method for the solution of the identity and transformation problems, which will also prove to be useful sometimes. If A is an element of the fundamental group, represented by an edge path $A_1 A_2$, then this uniquely determines a motion of the net which carries A_1 to A_2 . Because if there were a second motion, other than the rotation of the A-chain about the ideal midpoint, it would have to be a rotation about a real midpoint A' , and we have shown that such a motion of the net is impossible.   Now suppose that
the element A is transformed
[conjugated] by an arbitrary
element C .   We then find the
transformed element $B = C^{-1} A C$
by attaching to A_1 and A_2
the homologous edge paths
$A_1 C_1$ and $A_2 C_2$ which
correspond to the element C ,
and then shifting the initial

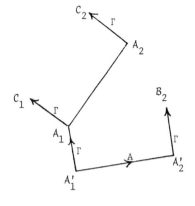

Figure 12

point C_1 of the path $C_1A_1A_2C_2$ to A_1 , giving the edge path A_1B_2 . Since A_1C_1 and A_2C_2 are homologous edge paths, the motion of the net corresponding to the element A simultaneously sends A_1 to A_2 and C_1 to C_2 . But A_1 is sent to A_2 by a translation V_a and C_1 is sent to C_2 by a translation $V_b = V_c^{-1} V_a V_c$, where V_c denotes the translation which sends A_1 to C_1 . Moreover, the translation V_b obviously sends the point A_1 to B_2 . We therefore have

(1) Two elements A and B of the fundamental group are transformable into each other [conjugate] when the corresponding translations V_a and V_b are transformable into each other, i.e. when

$$V_b = V_c^{-1} V_a V_c$$

Two mutually transformable [conjugate] translations V_a and V_b are distinguished by the fact that they represent the same motion of the net, i.e. they have the same ideal angle of rotation, and homologous ideal midpoints.

This theorem enables us to give a simple solution to the transformation problem for two translations. If X_1, X_2 and Y_1, Y_2 are pairs of points on the axis of the A-chain which correspond under rotations with the same angle, then $\overline{X_1X_2} = \overline{Y_1Y_2}$. Further, if Z_1 and Z_2 are two corresponding points on the axis of the B-chain, then $\overline{Z_1Z_2} = \overline{X_1X_2}$, since both segments correspond to the same angle and one axis is sent to the other by the motion V_c^{-1} . If we now denote the distance between two points which correspond under the rotation through the given angle as the

"<u>displacement length</u> σ_a <u>of the motion</u>", then we obtain the

theorem:

<u>A necessary condition for the transformability of two dis-</u>

<u>placements</u> V_a <u>and</u> V_b <u>into each other [conjugacy] is the</u>

<u>equality of their displacement lengths, i.e.</u> $\sigma_b = \sigma_a$.

However, this condition is not completely sufficient. In

order to find other conditions, we recall that our rotations about

homologous ideal points must carry the net onto itself. Thus if

we construct the image of the displacement segment $v_a = X_1X_2$ in

a single polygon, which is well known to be a sequence of individual

straight line segments, then the endpoints must coincide or else be

corresponding points on homologous boundary edges a_i, a_i^{-1} or

b_i,b_i^{-1} . Now the motion carries the axis segment X_1X_2 to X_2X_3 ,

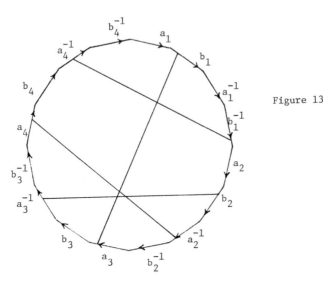

Figure 13

but since the net is unchanged in the process, X_2X_3 must map back

onto the same sequence of line segments in a polygon as X_1X_2 , and

indeed with the same initial and final point. Since the same is
true of the segments $X_3X_4, X_4X_5, \ldots, X_0X_1, X_{-1}X_0, \ldots$, we see that
the axis runs homologously through the different polygons. Thus
if $Y_{-1}, Y_0, Y_1, \ldots$ is another series of corresponding points on
the axis of the same motion, $\ldots, Y_{-1}Y_0, Y_0Y_1, Y_1Y_2, Y_2Y_3, \ldots$
always map onto the same sequence of segments in each polygon, as
do the segments $\ldots, X_{-1}X_0, X_0X_1, X_1X_2, X_2X_3, \ldots$, except that the two
images in general have different initial and final points. But
this means that the image of the segment $v_b = Z_1Z_2$ must also
coincide with the image of the segment $v_a = X_1X_2$, up to a cyclic
rearrangement of the sequence of segments, because the position of
this image is not altered by an arbitrary motion v_c^{-1}. Corresponding to each motion of the net we therefore have a uniquely determined cyclic sequence of line segments in a polygon, which we call
the "axis image of the motion". It then follows immediately that:
The necessary and sufficient condition for the mutual transformability [conjugacy] of two translations V_a and V_b is the
equality of their axis images.

We have therefore obtained a method for deciding, in a finite
number of steps, whether two motions, or the corresponding group
elements, are mutually transformable. Here one can also express
the condition for transformability analytically, and then arrive
at an analytic solution of the conjugacy problem.

5.   One-sided surfaces. This enables us to conclude our investigation of two-sided surfaces, and we shall now treat the one-sided
surfaces in a similar way. The best known example of such a surface, on which one can no longer distinguish two sides, is of

course the "Möbius band" which one can easily construct by pasting

together the two ends of a paper strip in such a way that one side

continues into the other. However, this figure is not the simplest

one-sided surface of all, because it still has a boundary curve.

In order to obtain a one-sided surface without boundary we divide

the boundary of a circular disc into four equal parts by four

points A, B, C, D, and identify the parts in such a way that A

coincides with B and C and D, in other words, AC is

defined as equal to BD, and CB as equal to DA. This diametric

relationship between the
boundary parts may be easily
visualised by suitably
joining together the identified parts. The result is
a closed one-sided surface
of the form T_1. The index of T denotes the
number k of circuit cuts
required to reduce a closed

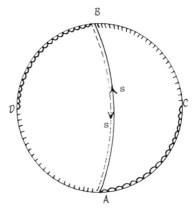

Figure 14

surface T_k to a disc; thus for two-sided surfaces k has the value

$2p$, i.e. twice the genus, and one also has the equation

$$k = \alpha_1 - \alpha_0 - \alpha_2 + 2 ,$$

where $\alpha_0, \alpha_1, \alpha_2$ denote the numbers of vertices, edges and faces

respectively in a polyhedron obtainable by continuous deformation

of the surface. In order to obtain the circuit cut for our

simplest one-sided surface T_1, we connect two homologous points,

say A and B of the disc boundary [Fig.14]. By the construction of the surface, this connection is a closed curve. All closed

curves on our one-sided surface are reducible either to a

point or to this one curve. The fundamental group of T_1,

i.e. the group of curves on this surface, has only a single

generator, which we will denote by s, and the relation $s^2 = 1$,

so that

$$s \quad \text{and} \quad s^2 (=1)$$

are the elements of this fundamental group.

A one-sided surface of higher connectivity can be produced

by cutting an ordinary two-sided ring [torus] or tube and pushing

one end through to the other in such a way that the outer side of

the tube joins up with the inner

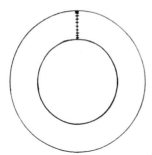 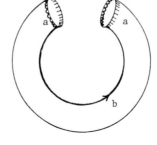

Figure 15

The resulting <u>one-sided ring surface</u> [Klein bottle] may be reduced

to a disc by two circuit cuts a and b, and hence it is a

surface of the form T_2. If the cut ring is spread out with a

definite direction of the cuts displayed then the result is a

rectangle $ABCD$ whose sides AB, BC, CD, DA are denoted in turn by

a, b, a, b^{-1}. In this single rectangle, as in the case of the two-

sided ring, curves on the surface
are represented by curve segments
with endpoints on the boundary of
the rectangle. In order to obtain
an unbroken image of a continuous
curve on the one-sided ring, we

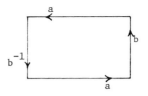

Figure 16

again replace the single rectangle by a net of rectangles which cover

the whole plane simply and without gaps, in which like-positioned or

homologous points of individual rectangles correspond to each other.

We can then regard this rectangular net as an infinite-sheeted

covering of the ring, the individual sheets of which meet along the

curves a and b . Now the rectangular net just constructed is at

the same time the group diagram of the fundamental group of the one-

sided ring, which is determined by the two generators a and b

and the relation

$$abab^{-1} = 1 .$$

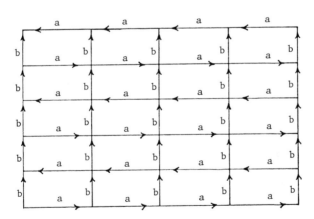

Figure 17

Namely, we have the four segments a, b, a^{-1}, b^{-1} issuing from each vertex of the net, the relation $abab^{-1} = 1$ is satisfied, and each relation of the group, i.e. each closed edge path of our net, is composed from this relation. Thus the construction of the group diagram gives an immediate solution of the identity problem. Two elements of the group are identical if and only if they are represented by edge paths with the same initial and final point; topologically, this means that a closed curve on the ring is contractible to a point if and only if its image in the net is closed.

While the identity problem is thus settled exactly as in the case of the simple two-sided ring, the solution of the transformation problem here takes a rather different form. This is because we have two different types of motion of the net. There are parallel displacements along the segments a and b, but a displacement along an odd number of b segments has to be composed with a reflection in the vertical. The first operation maps the horizontal lines into themselves, while the second, if it involves reflection, sends a horizontal line to one which is an odd number of horizontal strips away. Now since two group elements are mutually transformable if and only if the associated displacements can be transformed into each other, and two mutually transformable displacements are related via the same motion of the net, our problem is to bring the parallel displacements and reflections into a canonical form. Two pure parallel displacements P_c and $P_{c'}$, associated with the elements c and c', are obviously transformable into each other if and only if they have the same length

and direction, i.e. when they are identical. Thus the first condition for the mutual transformability of two elements c and c'
is in the theorem: two elements of the fundamental group are transformable into each other [conjugate] if b appears to an even
power in both elements, and the exponent sums of a and b in both
elements are respectively equal.

Two parallel displacements through an odd number of horizontal
strips, related via reflection in a vertical, are obviously transformable into each other only if they involve the same number of
horizontal strips; the reflection of the net can take place either
in a b-line or in the line connecting midpoints of opposite a-
edges. We thereby obtain a second condition for the mutual transformability of two elements in the theorem: two elements of the
fundamental group are transformable into each other [conjugate]
when both elements contain b to an odd power, in fact to the
same power in both, and the power of a is either even in both
or odd in both.

Thus, for example, the elements aab and b must be
transformable into each other; likewise the elements ab and
$a^{-1}b$. In fact, as one finds immediately with the help of the
group diagram,

$$a(b)a^{-1} = a(aba)a^{-1} = aab$$

and

$$a^{-1}(ab)a = a^{-1}(aba) = a^{-1}b ,$$

using the relation aba = b each time. Our two theorems now
yield the following canonical representation for the elements of
the group:

1. $a^n b^{2m}$,   2. ab^{2m+1},   3. b^{2m+1} ;

each element of the group is transformable into a unique element of
this form. This implies the theorem that each curve on the surface
is uniquely reducible to one of these three types. It may be remarked, incidentally, that the fundamental group of the one-sided
ring is not abelian, because, e.g., $aba^{-1}b^{-1}$ is not equal to 1,
but to a^2. However, our group does indeed have a non-trivial
abelian subgroup, namely the subgroup of all elements in which the
total power of b is even, and in fact this subgroup is isomorphic
to the group of the two-sided ring.

We obtain the next one-sided surface T_3 by cutting a
circular hole in the one-sided ring T_2 and identifying points
on the boundary of the hole diametrically, as with T_1. By
cutting along suitable circuits a,b,c and spreading it out,
we can reduce this surface to a regular hexagon with a,b,c,b^{-1},c^{-1},a
as its sequence of boundary sides. We now construct a net of
regular hexagons in the non-euclidean plane, with six such hexagons
meeting at each vertex, so that the hexagon angle is $\frac{2\pi}{6} = 60°$,
then we can label the sides of each hexagon in the sequence
a,b,c,b^{-1},c^{-1},a . This net of hexagons then represents, on

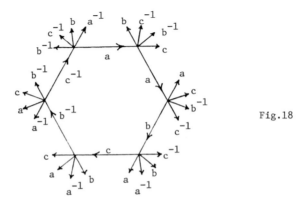

Fig.18

the one hand, the infinite-sheeted covering of our one-sided surface T_3 , but on the other hand it can also be regarded as the diagram of the fundamental group of our surface, which is given by the three generators a,b,c and the relation

$$bcb^{-1}c^{-1}a^2 = 1 .$$

The identity problem is again immediately solved by this group diagram, in the same way as in the earlier examples.

The transformation problem requires us to determine the mutually conjugate motions, and consequently to bring the motions of the net into canonical form. Here we again have to distinguish between two kinds of net motions. The first kind, which includes, e.g., the motions corresponding to the elements b and c , are the rotations about an ideal midpoint, already familiar to us. The second kind include, e.g. the motion determined by the element a , consisting of a reflection in the line associated with the rotation part of a , whose midpoint in our example is the centre of the polygon and hence finite. Our problem then is to find a canonical form for the second kind of motion. For this purpose we will determine a line in the net which is fixed by this kind of motion. If we drop the perpendicular MF from the rotation centre M to the reflection line and construct half the rotation angle, $\frac{\omega}{2}$, on each side of MF from M , then our motion which consists of rotation about M through the angle ω followed by reflection in the reflection line, obviously leaves invariant the perpendicular through F to the initial leg of the angle ω . We therefore call this the <u>reflection axis</u> of our motion. The motion of the net determines a certain length of displacement along the

reflection axis, called its

"displacement length". It

may now be easily shown that

the two motions are trans-

formable into each other

only if the corresponding

reflection axes have

homologous positions on the

net polygons and equal

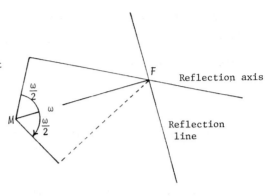

Figure 19

displacement lengths. We have then also solved the transformation

problem for the fundamental group of the T_3 .

Finally, we shall briefly treat the one-sided double ring T_4 .

Here, the fundamental group is given by four generators a_1, b_1, a_2, b_2

with the relation

$$a_1 b_1 a_1^{-1} b_1^{-1} a_2 b_2 a_2 b_2^{-1} = 1 .$$

In this case the group diagram is a net of regular octagons in the

non-euclidean plane, in which each octagon angle has the value

$\frac{2\pi}{8} = \frac{\pi}{4}$.   The identity problem is settled with the construction of

the group diagram, and the transformation problem is solved by

bringing the motions of the net into a canonical form.

6. Decomposable groups. In the groups treated previously we had to

satisfy only one relation, in which each generator appeared twice.

We then found that the group diagram was represented by a net of

regular polygons in the non-euclidean plane. The question now arises, whether each group with a single defining relation, containing each generator twice, has a group diagram in the form of a non-euclidean polygonal net. As we shall now see in a few examples this question must be answered in the negative.

Namely, let a,b,c be the generators of a group with the relation

$$abcabc = 1 .$$

We shall set abc = d and then take a,b,d as generators. The single relation which must then be satisfied is $d^2 = 1$, so that a and b are completely arbitrary. In order to find the group diagram for the group so determined, we only need five segments to issue from each vertex, corresponding to the operations $a, a^{-1}, b, b^{-1}, d = d^{-1}$, and the whole diagram contains no closed edge path. But this also shows that the diagram of our group

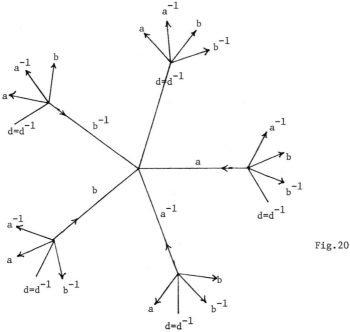

Fig. 20

cannot possibly be reduced to a polygon net. One does not see quite so immediately that the group with generators a,b,c,d,e and relation

$$a^{-1}cdc^{-1}bdb^{-1}ae^2 = 1$$

contains arbitrary elements. If we substitute in this

$$a^{-1}c = f \quad \text{and} \quad a^{-1}b = g,$$

then

$$c^{-1}b = f^{-1}a^{-1}b = f^{-1}g.$$

Hence if we consider b,d,e,f,g as generators of our group, then we now have to satisfy the relation

$$fdf^{-1}gdg^{-1}e^2 = 1,$$

so that the operation b is completely arbitrary. On this basis, the group diagram here can never be a net of polygons. We call a group which has a set of generators in which one or more are arbitrary a "decomposable group". For such decomposable groups, when there is only one relation and each generator appears twice, the group diagram is easy to construct, and the identity and transformation problems can then be solved.

7. Analytic remarks. For all indecomposable groups with only one relation, in which each generator appears twice, we have completely solved the identity and transformation problem, assuming that the group diagram is represented as a simple regular polygon net in the euclidean or non-euclidean plane. It may now be shown that these groups are isomorphic either to groups of linear nonhomogeneous transformations of two variables, as is the case, e.g.,

with the fundamental group of the two-sided and one-sided ring, which is represented in the euclidean plane, or to groups of linear homogeneous transformations of three variables leaving the equations $x_1^2 + x_2^2 = x_3^2 = 0$ invariant, when the group diagram is a regular polygon net in the non-euclidean plane. If we take the Poincaré representation of the non-euclidean plane as a basis, then the motions which come into consideration as group elements are analytically equivalent to the linear transformations

$$z' = \frac{az + b}{cz + d}$$

of the complex variable $z = x + iy$, where a, b, c, d are real numbers and the determinant $ad - bc > 0$. The proof of this assertion follows from the fact that our motions are circle-preserving mappings of the upper complex half plane, since the half-discs bounded by Poincaré lines, i.e. the non-euclidean half planes, are mapped onto each other and angles are preserved. Poincaré calls the groups of such real linear transformations of the complex variable "fuchsian groups". To obtain the fundamental groups of the one-sided surfaces, whose motions also include reflections, we must add to the above transformations the analytic expression

$$z' = -\bar{z} = -x + iy$$

for a reflection. We may call groups consisting of such transformations "extended fuchsian groups". Thus the fundamental groups of our surfaces are isomorphic to such extended fuchsian groups.

To round off this information we shall add a short remark on higher groups. As Poincaré has proved, the linear transformations of the complex variable $z = x + iy$ with complex coefficients $\alpha, \beta, \gamma, \delta$ are representable by motions of non-euclidean space. Poincaré's image of this space is a half space determined by the z-plane, the so-called "Poincaré half space", in which the non-euclidean planes are represented by spheres which cut the z-plane orthogonally. Since linear transformations of the z-plane send circles to circles, and hence at the same time can be considered to send orthogonal spheres to orthogonal spheres, a linear transformation of the z-plane corresponds to a motion of the non-euclidean space and conversely. The group of motions of non-euclidean space is identical with the group of linear transformations of z. This suggests going on to group diagrams in this non-euclidean space, for example in the case of the group with the generators a and b and the relation

$$aabba^{-1}b^{-1} = 1 .$$

Here we obtain the dual group diagram when we represent the four operations a, a^{-1}, b, b^{-1} by the reflections of a regular non-euclidean tetrahedron in its four faces. As a result of the relation between the generators, six tetrahedra must come together at each edge. The edge angle of the regular tetrahedron must therefore have the value 60°. But this is the case for the regular tetrahedron whose vertices lie on the absolute sphere which bounds the non-euclidean space in the Klein representation, because ordinary euclidean geometry holds at the points of the absolute sphere. By repeated reflection in

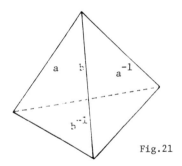

these tetrahedra we can produce
the whole group diagram. Each
element of the group may then
be associated with a particular
motion of the tetrahedral net.
Then by again bringing these

Fig.21

motions of the non-euclidean

space into a canonical form we obtain the solution of the identity

and transformation problems. Analytically, the motions are

equivalent to complex linear transformations of the complex

variable $z = x + iy$, because, as transformations of the Poincaré

half space, these motions have the property of sending spheres

orthogonal to the boundary plane to other such spheres, so that

the motion sends circles to circles in this boundary plane and

angles are preserved. It follows that here we are in fact dealing

with linear transformations of the complex variable z in the

boundary plane. Poincaré calls the groups of such transformations

"Kleinian groups" because Klein stimulated the development of these

ideas with his investigations of modular functions and automorphic

functions.

In his first major publication in topology, Dehn uses his ex-
perience in surface topology and group theory to formulate and
attack the major problems of 3-dimensional topology. Together
with Poincaré [1904], the paper which follows is the foundation
for almost all subsequent work in this field. A brief summary of
the contents will support this claim: statement of the word and
conjugacy problems for finitely presented groups, realisation of
such groups as fundamental groups of surface complexes, Dehn's
lemma, the diagram of the trefoil knot group, equivalence between
triviality of a knot and commutativity of its group, construction
of homology spheres by surgery, and proof that one of them (actually
Poincaré's homology sphere) has finite fundamental group; all making
their first appearance in print.

In Chapter I, §1, the word and conjugacy problems are formulated, and
Dehn   describes a process which determines the diagram of a finitely
presented group. He points out that the process is non-constructive,
however, being equivalent to solving the word problem, for which an
algorithm was then unknown (and now known to be non-existent). In
practice, Dehn only constructs diagrams whose correctness is immediate
from the defining relations, and his illustration is of the icosahedral
group, no doubt recalled from his lectures.

The short §2 of I represents finitely presented groups by
surface complexes; the generators by particular paths in a graph and
relations by discs spanning paths which represent the defining
relators. Later (in Chapter III, §1), Dehn observes that such a
representation of the fundamental group of a 3-manifold arises
naturally from its representation as a polyhedron with identified
faces. If one deletes the interior of the polyhedron and *then*
identifies the faces, the result is a surface complex (nowadays
called a *spine* of the manifold) with the same fundamental group.
The surface complex is useful in its own right for studying groups.
In particular, Reidemeister [1932] shows that subgroups correspond
to covering complexes, a result which can be used, e.g., to give an

elegant proof of the Kurosh subgroup theorem (Baer & Levi [1936]).

Chapter I, §3 states, but does not completely prove, the famous *Dehn's lemma*. A gap in the proof was found by Hellmuth Kneser, and he informed Dehn of it in a letter dated 22 April 1929. The interesting correspondence which ensued is preserved in the Humanities Library, Austin, Texas; it includes Kneser's elegant representation of Dehn's homology sphere (see later, and also Threlfall & Seifert [1930], p.66) as a dodecahedron with suitable identification of opposite faces. However, Dehn and Kneser were unable to fix the proof of the lemma, and a correct proof did not appear until Papakyriakopoulos [1957]. It may be significant that Papakyriakopoulos had previously undertaken a lengthy study of surface complexes (Papakyriakopoulos [1943]), which included proving the Hauptvermutung and solving the homeomorphism problem for them.

The lack of a proof of Dehn's lemma in 1929 was as big a disappointment to Kneser as to Dehn. Kneser's son, Martin, informed me in a letter that his father had planned to write a book on 3-dimensional topology based on Dehn's lemma. The collapse of Dehn's proof may have been instrumental in his moving from topology to several complex variables. Nevertheless, Dehn's "switchover" (Umschaltung) technique proved to be of fundamental importance when the theory of 3-manifolds finally took off in the 1960's, and Kneser's application of it, Kneser [1929], was then recognised as a decisive contribution (see e.g. Hempel [1976]).

Chapter II, on knots and their groups, begins in §1 by defining a knot to be non-trivial ("knotted") if it cannot be spanned by a non-singular disc. §2 encloses the knot by a tubular neighbourhood and observes that, even if the knot is non-trivial, there is always a curve on the tube surface which is null homologous in the complement of the tube. It follows that the homology of the knot complement can be killed by attaching a disc along this curve, and this is the basis of Dehn's homology sphere construction later. §3 arrives

at a presentation of the knot group, similar to the Wirtinger
presentation first explained in Tietze [1908], but by a different
route. Dehn appeals extensively to surface topology in his proof,
though this is not really necessary.

The first major result of Chapter II is in §4. Dehn's lemma
is used to show that a knot is trivial if its group is abelian
(the converse is obvious). He points out that one can then *decide*
whether a given knot K is trivial by construction of its group
diagram, since this in turn will decide whether the group is abelian.
(E.g., by checking whether $a_i a_j a_i^{-1} a_j^{-1} = 1$ for each pair of
generators a_i, a_j.) However, this is a little misleading, since
one needs a uniform algorithm which constructs the group diagram
from K , i.e. a uniform algorithm which solves the word problem
for knot groups, and such an algorithm was first given by Waldhausen
[1968]. By this time, a geometric algorithm for recognising the
trivial knot had already been given, by Haken [1961]. Haken's method
uses none other than Dehn's switchovers, in a very elaborate way, and
furthermore, Waldhausen's work is based on Haken's. One can therefore say that Dehn was on the right track, though he had a long way
to go.

Chapter II, §5 contains the diagram of the trefoil knot group,
and thereby solves its word problem. The diagram is non-planar and
quite unlike any group diagram previously constructed, though surprisingly simple and beautiful. Dehn uses it to construct the group
diagram for the homology sphere obtained by killing the homology of
the trefoil knot complement in an operation now known as Dehn surgery,
showing that for the simplest surgery the group is of order 120 and
in fact the extended icosahedral group he had studied in his lectures. It was shown by Weber & Seifert [1933] that Dehn's homology
sphere is the same as the first one discovered, by Poincaré [1904].
A simple pictorial proof is given in Rolfsen [1976]. Poincaré did
not show that its group was finite, only that it had the icosahedral
group as homomorphic image. This homology sphere is still the only

one known which has a finite non-trivial fundamental group.

It should be mentioned that Dehn had in fact introduced
surgery earlier, in the abstract Dehn [1907], for a different
homology sphere construction which is short enough to quote in
full:

> A very clear example of such a noteworthy manifold
> can be constructed as follows: a *knotted* torus T_2 in
> ordinary space divides the latter into a solid torus
> T_3 and a part K_3 not homeomorphic to it. Suppose that
> curves C resp. Γ are non-separating on T_2 and
> bounding in T_3 resp. K_3. One joins K_3 to a homeo-
> morphic body K_3' which is bounded by T_2' (with the
> curves C' and Γ') in such a way that T_2 is identi-
> fied with T_2', C' with Γ, and C with Γ'. In the
> resulting closed M_3, each curve is bounding when taken
> once. However, it is not homeomorphic to the ordinary
> space, since this M_3 is divided by the torus T_2 into
> two parts K_3 and K_3', neither of which is homeomorphic
> to a solid torus.

The geometric theorem used here by Dehn, that a torus in $\$^3$
bounds a solid torus on at least one side, was not proved until
much later, by Alexander [1924]. As I pointed out in Stillwell
[1979], one can instead justify the construction by a group
theoretic argument and Dehn's lemma. It is plausible that Dehn first
began thinking about the lemma in the context of this example. Nowadays, Dehn surgery is not used only to kill homology. It can in
fact be used to construct any orientable 3-manifold from $\$^3$
(Wallace [1960], Lickorish [1962]).

The final Chapter III derives representations of 3-manifolds by
polyhedral schemata and Heegaard decomposition, which of course
were already known, and then tries to see what light they throw on
the problems of recognising whether a manifold is homeomorphic to

s^3 , and settling the Poincaré [1904] conjecture that a simply connected closed 3-manifold is s^3 . Since these problems are still open, 70 years later, it is not surprising that Dehn's discussion of them is rather inconclusive.

For the sake of readability I have made the slightly anachronistic translations of "Elementarflächenstück" by "disc" or "2-cell" and "Elementarraumstück" by "3-cell" rather than the lengthy literal translations. I have also renumbered the first seven figures, which occur out of order in the original paper.

REFERENCES

.J.W. Alexander [1924]: On the subdivision of 3-space by a polyhedron. Proc. Nat. Acad. Sci. 10, 6-8.

R. Baer & F. Levi [1936]: Freie Produkte und ihre Untergruppen. Compositio Math. 3, 391-398.

M. Dehn [1907]: Berichtigender Zusatz zu III AB3 Analysis situs. Jber. Deutsch. Math. Verein. 16, 573.

W. Haken [1961]: Theorie der Normalflächen. Acta Math. 105, 245-375.

J. Hempel [1976]: *3-manifolds*. Ann. of Math. Studies 86, Princeton University Press.

H. Kneser [1929]: Geschlossene Flächen in dreidimensionalen Mannigfaltigkeiten. Jber. Deutsch. Math. Verein. 38, 248-260.

W.B.R. Lickorish [1962]: A representation of orientable combinatorial 3-manifolds. Ann. Math. 76, 531-540.

C.D. Papakyriakopoulos [1943]: A new proof of the invariance of the homology groups of a complex (Greek). Bull. Soc. Math. Grèce 22, 1-154.
[1957]: On Dehn's lemma and the asphericity of knots. Ann. Math. 66, 1-26.

H. Poincaré [1904]: Cinquième complément à l'analysis situs Rend. circ. mat. Palermo 18, 45-110.

K. Reidemeister [1932]: *Einführung in die kombinatorische Topologie*. Teubner, Leipzig.

D. Rolfsen [1976]: *Knots and Links*. Publish or Perish, Inc.

J.C. Stillwell [1979]: Letter to the Editor. Math. Intelligencer 1, 192.

W. Threlfall & H. Seifert [1930]: Topologische Untersuchungen der Diskontinuitätsbereiche endliche Bewegungsgruppen des dreidimensionales sphärischen Raumes I. Math. Ann. 104, 1-70.

H. Tietze [1908]: Über die topologischen Invarianten mehrdimensionaler Mannigfaltigkeiten. Monatsh. Math. Phys. 19, 1-118.

F. Waldhausen [1968]: The word problem in fundamental groups of sufficiently large irreducible 3-manifolds. Ann. Math. 88, 272-280.

A.H. Wallace [1960]: Modifications and cobounding manifolds. Can. J. Math. 12, 503-528.

C. Weber & H. Seifert [1933]: Die beiden Dodekaederräume. Math. Zeit. 37, 237-253.

ON THE TOPOLOGY OF THREE-DIMENSIONAL SPACE

In the present work, we topologically investigate closed curves (knots) in ordinary (i.e. hyperspherical) space (Chapter II), as well as three-dimensional space in general (Chapter III). Preceding these discussions are the necessary general considerations (Chapter I). In fact in §1 we consider groups of discrete operations, namely those constructed from a *finite* number of generating operations, among which a *finite* number of relations are given.

A diagram is constructed for such a group, a regular line segment complex, whose construction, in contrast to previous representations of general infinite groups, disposes of important problems connected with the group. At the same time this construction also involves some difficulties. In the course of this work group diagrams will be given for known (p.100) as well as for previously unknown finite and infinite groups (p.116). They are all connected with regular tessellations of euclidean or non-euclidean planes.

In §2 the construction of the fundamental group of an arbitrary surface complex follows. By means of the group diagram, each surface complex is associated with a regular line segment complex, infinite in general. In the case of closed two sided surfaces it is known that these give regular nets of 4n-sided cells in the hyperbolic plane, with 4n cells meeting at each vertex.

In §3 a general theorem on surface complexes is derived, the "lemma" which is often used in the later investigations. It is the 2-dimensional generalisation of the (trivial) theorem that one can connect any two points of a line segment complex by a polygonal path free of singularities.

In chapter II we first settle the definition of knot (§1) and in §2 introduce the enveloping tube of a knot. In §3 we construct the fundamental group of the knot, the definition of which is derived immediately in the simplest way from the plane projection of the knot. *If this group is abelian then the given curve is unknotted* (§4, Theorem 2). Further, a method is given by Theorem 1 for deciding whether a curve is knotted or not, by construction of the diagram of the fundamental group of the knot.

In §5, special knots are handled, in particular the trefoil knot and its relatives. The associated group diagrams are constructed, and in addition *Poincaré spaces* are derived from them, i.e. three dimensional manifolds which, despite being simply connected[*] and without torsion, are not homeomorphic to the usual space. The simplest one found here has a finite fundamental group and indeed it is the icosahedral group with reflections. The others have infinite groups. The group diagram is established for all of them; this settles the question whether a given curve of the manifold under consideration can be contracted to a point or not. This question had not previously been answered for such manifolds, nor did one know whether the associated fundamental group was finite or infinite. These results have the particular value of yielding a very simple method for the construction of infinitely many Poincaré spaces, of which the discoverer had constructed only one, in a complicated way.[**]

In §1 of chapter III different, easily visualised, methods for generating the three dimensional manifolds are given. For example, all two sided manifolds are divided into two pieces by a closed two sided surface, and the pieces are homeomorphic to the simplest part of space which can be bounded by such a surface. These methods of generation also lead to the construction of the fundamental group of the given manifold.

§2 deals with the important problem of the topological characterisation of ordinary space, without resolving the problem however. It treats the question of how ordinary space is to be topologically defined through the properties of its closed curves, and how to make

[*] In terms of homology (Translator's note).
[**] See Dehn, D. Math. Ver. 1907 p.573.

it possible to decide whether a given space is homeomorphic to ordinary space or not. The history of this problem began when first Heegard (Diss. Copenhagen 1898) and then Poincaré (Pal. Rend. v.13 and Lond. M.S. v.32) pointed out that in order to characterise ordinary space it does not suffice to assume that each curve bounds, possibly when multiply traversed. Indeed the manifolds with *torsion* exhibit this. Then Poincaré proved in Pal. Rend. 1904, by construction of a "Poincaré space" that it is even insufficient for each curve to bound when traversed once.

It now is natural to investigate whether it suffices to suppose that each curve of the space bounds on elementary piece of surface [disc]. This is also suggested at the end of Poincaré's work. However the reduction of the problem given in the foregoing work does not appear to lead directly to a solution. A deeper investigation of the fundamental groups of two sided closed surfaces seems to be unavoidable.

CHAPTER I Preliminaries

§1 Group theoretic aid (the group diagram)

In topology one is frequently led to problems of the following form:

A finite series of operations $a_1, a_2, \ldots, a_n$ are given as *generating operations* of a group. The group G is then completely determined by a finite number of *relations* between the defining operations, say of the form

$$A \begin{cases} \prod\limits_{i} a_{k_i^{(1)}}^{\varepsilon_i^{(1)}} = S_1 = 1 \\ \quad \cdot \\ \quad \cdot \\ \quad \cdot \\ \prod\limits_{i} a_{k_i^{(m)}}^{\varepsilon_i^{(m)}} = S_m = 1 \end{cases} \qquad (\varepsilon_i^{(\ell)} = +1 \text{ or } -1) \quad .$$

The group G is in fact completely determined by A. For if two operations S and T of the group are given as products of the a_i it is completely determined whether $S = T$ follows from the relations A or not, i.e. whether S = T is the case in the group with generating operations a_i and relations A , or not.

Our problems now run as follows:

1. A method is sought for deciding in a finite number of steps whether two products of the given operations a_i of G are equal or not, in particular whether one such product operation is equal to the identity.

2. A method is sought for deciding in a finite number of steps, given two substitutions S and T of G , whether there is a third, U , such that

$$S = UTU^{-1}$$

i.e. whether S may be shown by A to be a conjugate of T .

These two problems are completely solved for the special groups G_p given as follows:

$$G_p \begin{cases} \text{generating operations } a_1, b_1, \ldots, a_p, b_p \\ \text{relation: } a_1 b_1 a_1^{-1} b_1^{-1} \ldots a_p b_p a_p^{-1} b_p^{-1} = 1 \end{cases}$$

G_p is none other than the *fundamental group of curves on a two-sided surface* F_p of genus p . Our problem here is equivalent to the problem: when are two closed curves on F_p reducible to one another (with or without fixing a point)? The solution is obtained by a mapping of G onto a regular tessellation of the hyperbolic plane. We shall attempt the solution of the general problem through a generalisation suggested by this mapping.

We construct the following line segment complex C_1^G corresponding to a group G . Suppose

$$S_1 \equiv a_{k_1}^{\varepsilon_1} a_{k_2}^{\varepsilon_2} \ldots a_{k_\ell}^{\varepsilon_\ell} .$$

Then through a point Z we draw a circle (closed curve) K' with ℓ vertices $Z, P_1, \ldots, P_{\ell-1}$ and label ZP_1 with $+a_{k_1}$ or $-a_{k_1}, P_1 P_2$ with $+a_{k_2}$ or $-a_{k_2}$ etc., according as $\varepsilon_1, \varepsilon_2, \ldots$ are equal to $+1$ or -1. Then we draw a second circle K^2 through Z whose segments, starting from Z, are labelled by $+a_{k_2}$ or $-a_{k_2}, \ldots, +a_{k_\ell}$ or $-a_{k_\ell}$ (the signs again decided by the values ε_i). We proceed in this way until we have exhausted all cyclic transformations of S_1. There are then ℓ circles attached to Z, whose segments, taken in a suitable sense, have the given labelling. If they were traversed in the reverse sense, this labelling would be taken with the opposite signs. Now if two segments ZP and ZQ emanating from Z, running to P and Q respectively, carry the same labelling then P and Q are made to coincide, together with the segments themselves, and the process will only be continued as long as different segments emanate from Z. We also carry out the same process for the other vertices of the circle, again only as long as all segments emanating from a vertex are different. We call the line segment complex so constructed $\overline{C_1^1}$, and call Z its centre.

We now prove the following property of $\overline{C_1^1}$: If a closed curve of $\overline{C_1^1}$, traversed in a particular sense, consists of the segments $d_1, d_2, \ldots, d_q$ (all equal to $\pm a_i$) then by virtue of the relation $S_1 = 1$ the operation $d_1 d_2 \ldots d_q$ is likewise equal to 1: it results from the operation S_1 by conjugation and composition.

In fact: our assertion is correct for the initial state of C_1, where all closed curves of the complex go through Z and comprise curves of the form:

$$S_1, a_{k_1}^{-\varepsilon_1} S_1 a_{k_1}^{\varepsilon_1}, a_{k_2}^{-\varepsilon_2} a_{k_1}^{-\varepsilon_1} S_1 a_{k_1}^{\varepsilon_1} a_{k_2}^{\varepsilon_2}, \text{ etc.}$$

But each closed path whose initial and final point are Z is representable by a composition of these expressions. Any other choice of initial point however represents a cyclic interchange, and hence is obtainable by conjugation. We must now convince ourselves that this property cannot be lost with the coincidence of two segments originating from the same point. If $d = PQ = PR$ is such a segment we can assume that the curve in question goes through Q, since all other closed curves would also be closed before the

alteration and have the property claimed, by hypothesis. So suppose
k is a closed curve which goes through Q . Then the curve
$k' = dkd^{-1}$ was already closed before the alteration, and since by
assumption it results from S_1 by conjugation and composition, the
same is true of k , since it is a conjugate of k' .

We now introduce in succession the other relations $S_2, \ldots, S_m$,
where we can assume that each generating operation a_i appears in
at least one of them, or in S_1 . Otherwise we add to A the trivial
relation $a_i a_i^{-1} = 1$, without altering the group. We now apply again
the process described above until we come to a complex C_1^1 in which
no vertex has two identically labelled segments originating from it,
and each of the edges $a_1, \ldots, a_n$, $a_1^{-1}, \ldots, a_n^{-1}$ emanates from Z .

It is completely determined which of the curves S_1 ,

$$\alpha_{k_1}^{-\varepsilon_1} S_1 \alpha_{k_1}^{\varepsilon_1}, \ldots, S_2, \ldots, S_n, \ldots$$ goes through a point P of C_1^1 as
initial point. We now add to C_1^1 circles at each point P ,
labelled in such a way that now *all* curves of the collection

$$S_1, a_{k_1}^{-\varepsilon_1} S_1 a_{k_1}^{\varepsilon_1}, \ldots, S_n, \ldots$$ emanate from P . This is done for all

vertices of C_1^1 . We now modify the resulting complex by
identifying all edges which originate from a common point and are
identically labelled, thus obtaining a complex C_1^2 . Then we
construct C_1^3 from C_1^2 as we have constructed C_1^2 from C_1^1 .
In this way we obtain a sequence of line segment complexes which is
in general unbounded, though in certain circumstances it is bounded
(the series breaks off when all curves of the system $S_1, \ldots, S_n, \ldots$
already emanate from each point of a complex C_1^i).

(a) It follows as previously that if the segments of a
closed curve on C_1^i traverse $d_1, d_2, \ldots, d_q$ in order then the
operation

$$S = d_1 d_2 \ldots d_q$$

is generated from $S_1, S_2, \ldots, S_n$ by conjugation and composition,
i.e. S = 1 in G .

(b) We shall now show: Given an operation S, there exists a number r such that, when $i > r$, there is a curve in C_1^i with Z as initial point which represents S and is closed if S results from the S_i by conjugation and composition.

For let $S = d_1 d_2 \ldots d_f$ where the representation on the right hand side is chosen so that the generation of S by composition and conjugation of the S_i arises immediately, i.e. the right hand side consists of factors which are either *identically* S_i or terms which result from composition and conjugation of the S_i. One can then find r such that the curve corresponding to this representation (which is therefore comprised of the segments $d_1, d_2, \ldots, d_f$) originates from Z and exists in each C_1^i $(i > r)$. For a factor of the curve of the form originating from a point of C_1^r exists by construction in the C_1^i $(i > r)$ and returns to the same point. However, if a factor T of the curve is closed, then so too is the factor of the form UTU^{-1}, which lies in C_1^r, because U and U^{-1} are coincident but oppositely directed paths in C_1^i $(i > r)$, by our construction.

If S is originally given another form, then this results from the above by repeated deletions of two successive terms of the form e and e^{-1}. But such a deletion transforms a closed curve into another closed curve, as again follows immediately from the properties of our construction.

This yields: two distinct vertices P and U in a C_1^i can be brought into coincidence by extension of this complex. However, there is in any case a number r_i such that any two distinct vertices of C_1^i, which remain distinct in $C_1^{r_i}$, remain distinct in any C_1^ℓ: each curve of C_1^i which is unclosed in $C_1^{r_i}$ remains that way as the construction expands, and consequently represents an element of the group different from the identity. We shall denote the permanent form of C_1^i, as exhibited by $C_1^{r_i}$, by $[C_1^i]$, and construct the *infinite line segment complex* C_1^G, the *group diagram of the group* G, whose vertices and edges are obtained as vertices and edges of $[C_1^i]$ when we give i the successive values

1,2,...,n. We associate the centre Z with the *identity*, then each element $S = d_1 d_2 ... d_f$ is associated with the endpoint of a polygonal path. Two distinct elements receive distinct vertices of the diagram, and conversely, distinct vertices of the diagram represent distinct elements of the group. *The first fundamental problem is solved by construction of the group diagram* C_1^G. However the above gives a proof of the existence of the diagram, not a way of obtaining this diagram in a finite number of steps. We can express this more precisely as follows: in order to decide whether two elements S and T are equal we construct a C_1^i in which both elements are represented by polygonal paths originating at Z. If the endpoints of these paths are also distinct in $C_1^{r_i}$, then S and T are distinct elements of the group. We have certainly proved the existence of r_i, but we know no general method for really finding it. In the following section we shall derive a whole series of such group diagrams, which will be of the greatest use to us in the investigation of the corresponding groups and topological figures. Here we shall show that the group diagrams for the fundamental groups (defined at the beginning) of closed two-sided surfaces are quite simple for abelian groups, but this has no significance, because the solution of the fundamental problem for these groups is trivial.

As an example, the group diagram for the icosahedral group is given in Figure 1. The group is generated by the substitutions a_1 and a_2 with the three relations

$$a_1^5 = 1, \quad a_2^3 = 1, \quad a_1 a_2 a_1 a_2 = 1 .$$

In the figure the line segments, with sign indicated by the arrows, are numbered with the subscripts of the corresponding generating substitutions. One sees clearly how the 60 elements of the group are obtained from an initial element by a_1 and a_2.[*]

---

[*] The relation between this complex and the icosahedral group is known (see Maschke Am. Journ. 1896). Furthermore, the group diagram is not strictly new for *finite* groups. It is closely related to the Cayley "Colour diagram".

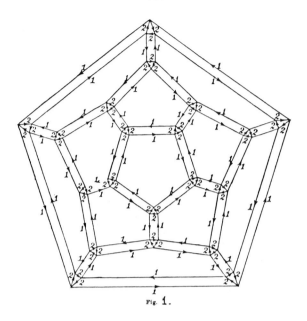

Fig. 1.

The group diagram for a finite group is a finite line segment complex (see the group diagram for the icosahedral group).

We shall draw a general conclusion from our construction: we saw that each closed curve of the diagram represented an operation equal to the identity because it resulted from the S_i by composition and conjugation. On the other hand, each operation equal to the identity is represented by a closed curve. Thus we have the theorem:

Each operation equal to the identity in a group given by generators and defining relations $S_1 = S_2 = \ldots = S_m = 1$, *results from the* S_i *by composition and conjugation.*

§2. The fundamental group of a surface complex.

In order to employ the properties of infinite groups in topology, we need the following observations.

1. *In a line segment complex* C_1 *with* α_0 *vertices and* α_1 *edges one can draw* $\alpha_1 - \alpha_0 + 1 = \mu$ *circles (closed polygonal paths)* $\alpha_1, \alpha_2, \ldots, \alpha_\mu$ *which go through a point* 0 *of* C_1 *which is chosen*

so that each other circle of C_1 *can be continuously transformed in* C_1 *into a circle* k' *which consists of circles* $\alpha_1, \ldots, \alpha_\mu$ *suitably ordered and directed.*

We write

$$k = k' = \alpha_{k_1}^{\varepsilon_1} \ldots \alpha_{k_2}^{\varepsilon_2} \ldots \alpha_{k_n}^{\varepsilon_\mu} \quad \left(\begin{array}{l} \varepsilon_i = +1 \text{ or } -1 \\[4pt] k_i = 1, 2, \ldots \text{ or } \mu \end{array} \right)$$

to denote that k' results from running through first α_{k_1}, then $\alpha_{k_2}, \ldots$, finally α_{k_n}, and in positive or negative direction according as $\varepsilon_1, \ldots, \varepsilon_n$ are $+1$ or $=1$.

The proof of the theorem comes immediately from known theorems on the reduction of curves on closed two sided surfaces when one considers the two-side surface of genus $p = \mu$ which results from C_1 as follows: one replaces each n-tuple node by a sheet with $n - 1$ holes [i.e. with n boundary curves (Translator's note)] and each line segment by a pair of such holes [boundary curves] and a tube connecting them.

2. We convert C_1 into a surface complex C_2 by putting discs through m circles $k_1, k_2, \ldots, k_m$.

Let $k_1, k_2, \ldots$ have the representation

$$\left. \begin{array}{l} k_1 = a_{k_1^{(1)}}^{\varepsilon_1^{(1)}} \, a_{k_2^{(1)}}^{\varepsilon_2(1)} \ldots a_{k_{n_1}^{(1)}}^{\varepsilon_{n_1}^{(1)}} \\[20pt] \qquad \cdots \\[12pt] k_m = a_{k_1^{(m)}}^{\varepsilon_1^{(m)}} \ldots a_{k_{n_m}^{(m)}}^{\varepsilon_{n_1}^{(m)}} \end{array} \right\} A \; .$$

Then we construct the group whose operations are $a_1, \ldots, a_\mu$ and whose relations come from replacing the left hand side of A by 1's. This group is called the *fundamental group* G_{C_2} of the *surface complex* C_2. *Each circle* k *of* C_2 *corresponds to an operation of this group through its representation by the* a_i *and conversely, each operation of the group is a circle of* C_2.

When one notes (a) that each operation of G_{C_2} that is equal to 1 results from the elements of the group corresponding to $k_1, \ldots, k_m$ by composition and conjugation, (b) that each of collection of surface pieces comprising a disc includes at least one whose removal leaves a disc remaining, it follows easily that:

The necessary and sufficient condition for a circle k *in* C_2 *to bound a disc is that the operation of the fundamental group corresponding to* k *be equal to 1.*

§3.   A topological aid (the "Lemma").

We shall often need the following theorem from the topology of surface complexes, which we shall refer to as the *lemma*, because of its important place in this work.

Let C_2 *be a surface complex in the interior of a homogeneous manifold* M_n *(n > 2). Let* k *be a curve in* C_2 *which bounds a (singular) disc* E_2'. *If* E_2' *has no singularities on its boundary, then* k *also bounds a completely non-singular disc in* M_n.

This theorem is obvious when the M_n has more than three dimensions, for in such a manifold each two dimensional figure has a neighbour without singularities. We can therefore assume $n = 3$.

1(a)   Let AB be a multiply counted edge $(A \neq B)$, at which a number of sheets meet, i.e. AB bounds some pairs of 2-cells whose preimages form adjacent pairs of 2-cells in the non-singular preimage E_2^0 of E_2^1. Since we are inside a homogeneous M_3, by hypothesis, the neighbourhood of AB in M_3 is a 3-cell, and we can therefore speak of the sheets intersecting (cutting) or touching along AB.

We now assume that all the sheets along AB *touch,* so that
there is a sheet with no others on one side of it, and we can replace this sheet by another which lies on this side and contains
no interior point of the edge AB , without creating any new intersections with the other sheets. We can then proceed in the same
way with the other sheets along AB , until sheets no longer touch
along AB , so that AB ceases to be a singular edge, and no new
singularities appear.

(b) Let A be a multiply counted vertex at which a number of
sheets meet, with the latter having no common edge emanating from
A . Since the neighbourhood of A in M_3 is again a 3-cell, we
can find a sheet which has no others to one side of it, and therefore replace it by a sheet on this side which no longer goes
through A , without introducing any new intersections with the
other sheets. Proceeding in this way, the singularity at A can
be removed, without introducing new singularities.

We see from (a) and (b) that we can replace an E_2' with
singularities by a non-singular E_2 , provided that only touching
sheets occur at singular edges. Our theorem will therefore be
proved when we give a process which transforms a given E_2' into
one whose singularities are all of the above type.

2(a) First, one can remove all n-tuply (n > 2) counted
singular edges by replacing individual sheets by neighbours in
the way one resolves on n-tuple point of a curve into double points.

(b) By modifications of type (1) we remove all touching of
sheets at vertices or edges. This can result in other singular
points, at which two sheets multiply intersect (i.e. there can
be lines of intersection with corresponding multiple points).
Such occurrences are removed by replacing these sheets by neighbours.
Finally, by the same process, we can resolve each n-tuple point
(n > 3) into triple points. These transformations make *each
singular point*: either a general point on a singular double line,
or an ordinary triple point through which three sheets pass, like
the three cartesian coordinate planes at the origin ("branch points"

are not singular points in our terms, since only one [self-inter-secting] sheet goes through such a point). The *singular lines* on E_2' are only: 1. Pairs of singular segments with common endpoints, which constitute unclosed double lines ending at simple branch points, 2. closed lines which constitute closed double lines, either cyclically or pairwise. No single edge (of the non-singular preimage) can end at a single point of E_2' unless another singular edge emanates from it. Otherwise this point would lie on the boundary of E_2' contrary to the hypothesis: *the boundary would meet an interior point of* E_2'.

(A) We shall first deal with *unclosed double lines*. Let ℓ' and ℓ'' be two edges on the non-singular preimage E_2^0 of E_2' with the same initial and final vertices A and B. The lines ℓ' and ℓ'' coincide in E_2' and constitute a double line ℓ. A and B are the corresponding branch points. ℓ' and ℓ'' can have double points and also intersect each other. Let C be such a point of intersection, so that $C = C'$ on ℓ' corresponds to say C'' on ℓ'', and we can assume C' and C'' are distinct, otherwise C could be regarded as the end and branch point of the double line (see Figure 2).

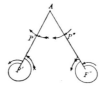

Figure 2   Figure 3

Now let AF' and AF'' (see Figure 3) be two corresponding non-singular segments of ℓ' and ℓ'' without a common point; then we can make a *"switchover"* along the corresponding segment AF of ℓ, i.e. join the half sheets of the sheets which meet along AF in another way, indeed, so that the new sheets touch along AF and the neighbourhood of A is two-sheeted. This can be done by connecting each half sheet with that half of the other sheet, which is not separated from it in E_2^0 by either of the lines AF' or AF" (see Figure 3, where this joining is indicated by different types of arrows). As a result, the point A, which was previously counted singly, becomes double; the double point F becomes single. Thus the number α_0 of vertices of E_2' remains unchanged, and hence also its characteristic, and since the new surface does not fall apart, as e.g. a glance at the figure shows, it remains *simply* connected. From the point F there is another unclosed double line k, which however is not necessarily identical to a part of ℓ. Since the intersection points of the segments constituting the double line do not themselves correspond, we can repeat this process until only a segment DE = {(DE)',(DE)"} of the double line remains, on which there is no intersection point of the pieces (DE)' and (DE)" and no double point of them. (DE)' and (DE)" are thus non-singular segments without common points.

It is now easy to see that we can cut sheets along this double line, by switchover of sheet halves, so that the neighbourhoods of D and E remain *one*-sheeted. The sheets to be joined are indicated by like shading in Figure 4. Again the number of vertices is unchanged, the new surface is connected, and hence *simply* connected. Thus if we remove the touching along AB, using (1), we have removed the singularity of ℓ without introducing new singularities or changing the connectivity. In this way we can replace E_2' by a singular disc E_2'' with the same boundary curve, in which the only singularities are closed double lines.

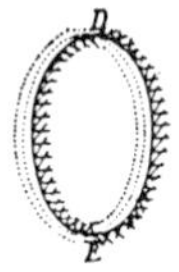

Figure 4

(B) It still remains to remove closed double lines. We collect them into groups, where a group consists of the segments in a double line together with all segments of the singular lines through points of the first double line, plus all segments of singular lines through singular points of the latter family of lines, etc.. The segments of a group thus constitute a series of closed curves on the non-singular preimage of the disc. Thus if we have any two points of the disc which do not lie on these lines, then all their connecting segments have either an even or odd number of points in common with each line. We now take any pair of corresponding segments of the lines in a group and switch over the half sheets along them in such a way that half sheets are connected when any two of their points have a connecting segment which meets the lines of the group in an odd number of points. We now claim the following: when we continue each switch-over sufficiently far along one side of the segment, we come to all segments of the group, with the exception of a pair of corresponding segments on which no singular point lies. The disc becomes a two- or one-sided surface with the number p = 1 resp. k = 2 , on which the remaining segments constitute a closed, non-singular, non-separating, two-sided curve. If we cut the surface along it and join the segments of the two boundary curves together, then the switchover is completed for all lines of the group, i.e. without introducing new singularities we have converted the disc into a new, simply connected manifold in which the singularities along the lines of the group can be removed by 1. Thus the lemma will be completely proved by a proof of the above assertion.

1. We show that continuation of the switchover leads to all segments of the group: if we continue the switchover through the next singular point, then we obviously obtain a connected manifold, in which the outgoing singular line from the endpoint of the switchover has changed: before the switchover it ran a second time through the singular point. Then, as a result of the switchover, the two branches through the singular point other than the switchover segment are them- selves switched over, as is shown in Figure 5 by the dotted lines. Thus if we now come from the endpoint of the switchover segment back to the singular point we must

Figure 5

again remove ourselves from it on the other branch. This other branch belongs to a second double line, belonging to the same group. By traversing the latter we again come back to the singular point (N.B. the line is indeed closed) and depart from the point on the segment still remaining, which belongs to the first singular line. If we follow this further, we come back to the initial point of the switchover segment. Thus we see: the singular line running from the initial point of the switchover segment consists of the remaining part of the first singular line, and the other line which emanates from the singular point, and for this new line the point in question is no longer singular. Proceeding in this way, we see that by continuation of the switchover and running through all lines of the group we finally obtain the singular line emanating from the endpoint of the switchover as a segment, ending at the initial point, which has more singular points. New singular segments do not occur.

2. The surface obviously remains connected under the switchover. The multiplicity of all vertices remains the same, except at the initial and final point of the switchover, which were double and have now become single. Thus the new surface has the number $p = 1$ resp. $k = 2$.

3. Suppose $A'B'$ and $A''B''$ (see Figure 6) are the nonsingular segments along which we have begun the switchover. But the latter is directed in such a way that we can connect two points P' and P'' on the sides of $B'A'$ and $B''A''$ which go into each other by a polygonal path Π which has an odd number of points in common with the edges of the group which remain singular. If we now continue

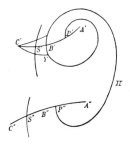

Figure 6.

the switchover from the next singular vertex S say to (C',C'') , then we can find a curve Π' which also has an odd number of points in common with the remaining edges of the group; we need only replace each segment XY of Π cutting B'C' or B"C" by the segment pair YC',C'X resp. YC", C"X, which has three points in common with each segment. The result of this process on Π is again a *closed* curve with the desired property. Proceeding in this way we finally are left with just two non-singular segments AZ of the group, and we can find a closed curve Π⁰ which has an odd number of points in common with both these segments. These therefore constitute a *non-separating* curve on the surface.

4. We have finally to prove that the curve consisting of the two segments AZ is two-sided. We consider the curve consisting of the two segments A'B' and A"B" with which we began the switchover. This curve is two-sided, its two sides consist of the two sides of A'B' resp. A"B".[*] Thus nothing is changed when we continue the switchover to the point C over a singular point. But the curve (A'B',A"B") resp. (A'C',A"C") is non-separating, because there is a closed curve Π, resp. Π', which meets it in only a single point. If we continue the switchover up to the point Z , we therefore obtain a two-sided, non-separating simple curve which is *cut* by the curve formed by the two segments AZ in two points, namely A and Z . Consequently this latter curve is also two-sided, which completes our proof.

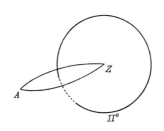

Figure 7

COROLLARY. We shall use the lemma in the following form:

K also bounds a non-singular disc under hypotheses which are the same as above except that singularities do lie upon K in the E_2' given originally, but K does not intersect the interior of E_2' (i.e. does not pass from one side of E_2' to the other).

---

[*] The two singular segments emanating from A lie on *different* sides of the curve (A'B',A"B").

In fact this hypothesis suffices for what is necessary to the proof (see p.104)- the impossibility of singular segments which terminate in the interior of E_2^0 without being attached to another singular segment .

CHAPTER II Knots and groups
§1 Definition

If two figures G and G' of ordinary space transform into each other by a continuous deformation of the whole space then in the encyclopedia article G and G' are called *iso-topic* in relation to the space. Two non-singular line segments, discs, spheres or 3-cells [in ordinary space] are isotopic to each other. On the other hand there are homeomorphic, but non-isotopic, closed curves, bounded or closed surfaces of higher genus: let U be the boundary curve of a non-singular disc, K another closed space curve without singularities. Then K is isotopic to U iff there is a non-singular disc bounded by K . The two are not iso-topic in general, namely when K is a knotted curve in the usual sense.

Definition: *A closed (non-singular) curve K is called un-knotted iff it is isotopic to the boundary of a non-singular disc.*

In the third paragraph of this chapter we shall give a method of deciding, for each topologically defined space curve, whether it is knotted or not.

While the use of infinite groups is necessary later, we shall first of all develop purely geometric properties of knots, which are also important in what follows.

§2 The knotted tube

The boundary of the neighbourhood of a closed space curve K is a ring-surface R (a tube). We find it as the boundary of a domain J consisting of two 3-cells E_3^1 and E_3^2 : E_3^1 and E_3^2 have two discs in common, E_2^1 and E_2^2 , and the space curve K consists of two segments which connect two points of E_2^1 and E_2^2 through the interiors of E_3^1 and E_3^2 . We notice that each nonself-intersecting curve on R , which intersects the boundary of E_2^1 and E_2^2 only once (it does not separate R) bounds, to-gether with K , a singularity-free strip in J and hence is isotopic to K in J . Thus if K is knotted, so is each such curve. The domain which, together with the interior J of the tube, makes up the whole space, will be called the complement space A ; further, the boundary of E_2^1 , which by construction bounds a simply connected piece of surface in J , will be denoted by $\widetilde{\mathcal{J}}$. We shall show that among the curves which cut $\widetilde{\mathcal{J}}$ once there is one which is bounding in A (null homologous). In fact: if we add to A the 3-cell E_3^1 then A becomes a 3-cell, in which each closed curve is null homologous. Thus each closed curve of A is homologous to a number of curves on the boundary of E_3^1. However, since only the boundaries of E_2^1 and E_2^2 are not bounding in A , and both are homologous to $\widetilde{\mathcal{J}}$, it follows that each curve in A and on its boundary is null homologous in conjunction with a multiple of $\widetilde{\mathcal{J}}$. Thus if C is an arbitrary curve on the tube cutting $\widetilde{\mathcal{J}}$ once, then in A

$$C \sim n\widetilde{\mathcal{J}}$$

so $$C - n\widetilde{\mathcal{J}} \sim 0 .$$

But one can always find a singularity-free curve $\mathcal{U}$, cutting $\widetilde{\mathcal{J}}$ once, which is homologous to $C - n\widetilde{\mathcal{J}}$. It follows, then, that

$$\mathcal{U} \sim 0$$

in the complement space A , as claimed. Indeed, the piece of surface bounded by $\mathcal{U}$ in A can be simply connected iff K is unknotted. For in this case, by addition of the singularity-free strip described above, we can obtain a disc bounded by K without boundary singularities and hence, by the lemma, without singularities at all, so that K is unknotted. We shall come back to this in the next paragraph.

§3 Construction of the fundamental group of a knot

Each line segment complex yields (see Chapter I, §2) a multiply connected surface: the segments are surrounded by tubes which connect the vertices. We shall consider these surfaces specially for the plane projection of a knot. E.g. the simplest projection of the simplest knot, the trefoil knot, yields a surface of genus 4. In general a projection with $n - 1$ crossings yields a surface F of genus $p = n$.

In order to have a definite intuition as a basis for the following discussion we think of the surface F symmetrically in relation to the plane of projection. In addition it can be placed so that it has $n + 1$ curves, in common with the plane of projection: the outer contour, and the contours of the n compartments. We transform the latter curves on F by double segments, from a common point P on (say) the upper side of the surface to points on the inner contours and back again, which are disjoint from each other and the contours except at these points. We denote the n curves so constructed through P on the surface by $C_{n+1}, C_{n+2}, \ldots, C_{2n}$. We put another n curves $C_1, \ldots, C_n$ through P so that C_i *cuts* the curve C_{n+i} in P (i.e. goes from one side of C_{n+i} to the other), C_i does not cut the curve C_{n+h} ($h \neq i$) and two curves C_i and C_k have no point in common outside of P. Then the surface is modified into a polygon by cutting along $C_1, \ldots, C_{2n}$, the boundary of which consists of the curves $C_1, C_{n+1}, C_1^{-1}, C_{n+1}^{-1}, \ldots$, the notation indicating the given direction of passage. Each curve L on F is equivalent to a number of curves C_i in a definite order, as shown by an expression

$$C_{\ell_1}^{\varepsilon_1} C_{\ell_2}^{\varepsilon_2} \ldots C_{\ell_m}^{\varepsilon_m} .$$

If we construct the infinite (fuchsian) group with the generating operations $C_1, C_2, \ldots, C_{2n}$ and the single relation

$$C_1 C_{n+1} C_1^{-1} C_{n+1}^{-1} = 1$$

then each element of the group corresponds to a curve on F and conversely; further, each element of the group capable of being proved = 1 from the relation corresponds to a curve bounding a simply connected piece of the surface and conversely.

Now if we construct a new group with the same generating operations and

$$C_{n+1} = C_{n+2} = \cdots = C_{2n} = 1$$

then the above relation follows and further, each curve on F whose corresponding substitution in the new group = 1 bounds a disc in the complement space of F, since this is the case for $C_{n+1}, \ldots, C_{2n}$. Conversely also, each surface curve L which bounds a disc in the complement space corresponds to a substitution of the group which is equal to 1. In order to see this one constructs a non-singular disc in the complement space bounded by $C_{n+1}, \ldots, C_{2n}$. If this is taken double, then together with F it constitutes a sphere. A disc E, whose boundary is L, and which we can assume singularity-free by the lemma, intersects this sphere, if at all, in closed curves. Each of the pieces into which E is decomposed may be replaced by a piece of the sphere and thus results in a disc bounded by L which is comprised merely of pieces of a sphere. However these are bounded by the curves $C_{n+1}, \ldots, C_{2n}$ or the curves $C_1 \ C_{n+1} \ C_1^{-1} \ C_{n+1}^{-1} \ldots$, from which it follows that the element of the group corresponding to L is equal to 1. Since we can omit generators which are equal to 1 from a group, we obtain the curves on the surface F in relation to the complement space as those generated by $C_1, \ldots, C_n$, between which there are no relations.

We shall now add to the complement space of F n - 1 3-cells, corresponding to the n - 1 crossings of the projection figure: Let ABCD (see Figure 8) be a curve around one of the crossing points on the surface, and indeed let the points A,B,C,D lie in such compartments as correspond to a branch through the *under* side of

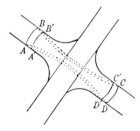

Figure 8

the crossing. Further, the segments AB and CD shall lie on the upper side of F, BC and AD on the lower side. Let A'B'C'D' be a neighbouring curve of ABCD on F. Then we remove from the interior of F a 3-cell bounded by the strip between them on F and two discs bounded by them inside F and add it to the complement space. We proceed similarly with each other crossing. We then obtain a complement space bounded by a ring surface R knotted in exactly the same way as the original curve K.

The n - 1 curves ABCD etc. are in the complement space of F, and hence reducible to curves of the form

$$C_{\ell_1}^{\varepsilon_1} C_{\ell_2}^{\varepsilon_2} \cdots, \quad (\varepsilon_k = \pm 1, \ \ell_i = 1, 2, \ldots, \text{ or } n)$$

as shown above. We shall denote the n - 1 curves in this form by $S_1, \ldots, S_{n-1}$.

In relation to the complement space A of R the curves of R then have the group:

$$G_K \begin{cases} \text{generating elements: } C_1, C_2, \ldots, C_n \\ \text{relations: } S_1 = S_2 = \ldots = S_n = 1. \end{cases}$$

This group G_K will be called the fundamental group of the closed space curve K.

We shall now explain how one may obtain the n - 1 relations in the simplest way directly from the projection figure: let (Figure 9) m_1, m_2, m_3, m_4 be the numbers of the compartments which meet at the crossing point, in the order $m_1 m_2 m_3 m_4$ around it. Suppose now that m_1 and m_4, likewise m_2 and m_3, meet along the lower branch, then *the relation corresponding to the crossing m*

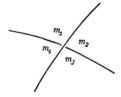

Figure 9

is

$$S_m = C_{m_1} \, C_{m_2}^{-1} \, C_{m_3} \, C_{m_4}^{-1} = 1 \, ,$$

when the sense of the C_i is suitably chosen. For the crossings which occur at the outer region the term corresponding to this region is simply left out. The proof of this assertion is easy with the help of theorems on the reduction of curves on surfaces.

All relations of the group G_K are thus three- or four-termed. In this way we obtain the fundamental group of the trefoil knot (Figure 10):

$$\begin{cases} \text{generating elements:} & C_1, C_2, C_3, C_4 \\ \text{relations} \begin{cases} C_1 \, C_4^{-1} \, C_2 = C_2 \, C_4^{-1} \, C_3 \\ \qquad\qquad = C_3 \, C_4^{-1} C_1 = 1 \end{cases} \end{cases}$$

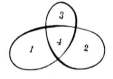

Figure 10

§4 Unknotted space curves

An unknotted space curve may be characterised as the boundary of a non-singular disc. Since the ring surface R is knotted like the given curve K, the curve $\mathcal{U}$ on R which is bounding in the complement space is knotted or unknotted according as K is. If K is unknotted then $\mathcal{U}$ bounds a non-singular disc in the complement space A. Thus if $S_\mathcal{U}$ is the substitution representing $\mathcal{U}$ in the group G_K, $S_\mathcal{U}$ must $= 1$. Conversely: if $S_\mathcal{U} = 1$ then in the complement space $\mathcal{U}$ bounds a disc which, since $\mathcal{U}$ lies on the boundary of A, can have no essential boundary singularities. Consequently $\mathcal{U}$ also bounds a non-singular disc in A, by the lemma. Then we have the theorem.

THEOREM 1: *For a space curve K to be unknotted it is necessary and sufficient that a certain substitution $S_\mathcal{U}$ equal 1 in the group G_K corresponding to K.*

In the case that K is unknotted, and hence also the tube R, all curves on R are reducible in the complement space to powers of a single curve which loops around the bounding curve $\widetilde{J}$ on R. So in this case G_K is isomorphic to the (abelian) group, generated by a single operation, without relations. Conversely, if G_K is an abelian group, it follows that K is unknotted. For since $\mathcal{U}$ is null homologous in the complement space it follows that

$$\mathcal{U} \equiv \sum_1^n \nu_i \, c_i \equiv \sum_1^{n-1} \lambda_j \, s_j$$

where s_j stands in place of the curve on R corresponding to S_j. But it follows from this congruence, if changing the order of substitutions is allowed, that

$$S_{\mathcal{U}} = 1$$

so that the sufficient condition of Theorem 1 has been reached. We therefore have

THEOREM 2: *K is unknotted iff the fundamental group of K is abelian.*

If a knot group is abelian, then it is isomorphic to the group $\{s^\alpha\}$.

One can apply yet another method for recognising unknotted curves: let $\mathcal{R}$ be any singularity-free curve on the ring surface R satisfying the homology

$$\mathcal{R} \sim \widetilde{J} + m\mathcal{U} \ .$$

(Each curve $\mathcal{R}$ which cuts $\mathcal{U}$ only once satisfies a homology of this type.) $\widetilde{J}$ (the curve bounding in the interior) can be represented by substitution C_i (where i is the index of a compartment which has a segment in common with the outer boundary). Let $\mathcal{R}$ be represented by the substitution $S_{\mathcal{R}}$ of G_K. Now if G_K is abelian, then

$$S_{\mathcal{U}} = 1 \quad \text{and} \quad S_{\mathcal{R}} = C_i .$$

If one now adds to A a slice of J bounded by a strip lying along $\mathcal{J}$, then A becomes a 3-cell for any knot. However the group of the curves on R relative to this 3-cell consists of G_K with the additional relation $C_i = 1$. But since all curves on R bound a disc in this 3-cell the group which results from G_K by addition of $C_i = 1$ must be the identity. Consequently G_K must also become the identity, when K is unknotted, when one adds $S_{\mathcal{R}} = 1$. However if K is knotted, this will not be the case in general, as we shall show in the next paragraph. On the contrary, knotted tubes give rise to manifolds of the kind discovered by Poincaré (Pal. Rend. 1904), and which we shall therefore call *Poincaré spaces*. These are [homologically] simply connected, closed M_3 without torsion, i.e. each closed curve is bounding in M_3 when traversed once. But they are in general not homeomorphic to ordinary space. Such a manifold cannot, in this general case, have the identity as fundamental group, as will be proved in the last chapter. Consequently we have (see construction of the fundamental group in M_3, Chapter III, §3).

K is knotted if addition of any one of certain relations $S_{\mathcal{R}} = 1$ to the fundamental group G_K does not result in the identity.

§5 Special knots and Poincaré spaces

    (a) The trefoil knot

    As we showed above, the fundamental group of the trefoil knot is:

$$\begin{cases} \text{generating substitutions:} \quad C_1, C_2, C_3, C_4 \\[2mm] \text{relations:} \quad C_1\,C_4^{-1}\,C_2 = C_2\,C_4^{-1}\,C_3 = C_3\,C_4^{-1}\,C_1 = 1 \end{cases}$$

The *group diagram* (see Chapter I, §1) is placed on a strip (Figure 11) bounded by lines consisting of segments C_4. The two parallel chains of segments C_4 are now connected by a polygonal path $C_3 C_2 C_1 C_3 C_2 C_1$ etc. so that the vertices lie alternately on the two C_4-chains. In this way we obtain three kinds of vertex on the strip:

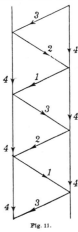

Fig. 11.

where C_1 and C_2^{-1} go out, where C_2 and C_3^{-1} go out, and finally where C_3 and C_1^{-1} go out. We now attach to the boundary of this strip two more strips, running in the same direction, so that all three types of vertex come together at each vertex of the common boundary. We only need to make sure that this is the case for a single vertex, in order to obtain it for all vertices. At each of the three free boundaries of this triple strip we attach two further strips in the same way, and so on. We easily see that the resulting infinite line segment complex is the group diagram of the given group. For emanating from each point we have eight segments

$$C_1, C_2, C_3, C_4, C_1^{-1}, C_2^{-1}, C_3^{-1}, C_4^{-1}$$

and hanging from each of these segments are the two resp. three circles which correspond to the two resp. three relations in which the corresponding substitution appears.

We now see immediately that the group is not isomorphic to the group $\{s^\alpha\}$, so that it is not abelian. For example the polygonal path $C_1 \, C_4 \, C_1^{-1} \, C_4^{-1}$ is not closed.

The curve $\mathcal{U}$ on R corresponds to the substitution $S_{\mathcal{U}} = C_1 \, C_2 \, C_3 \, C_2^{-1} \, C_2^{-1} \, C_2^{-1}$. We see immediately from the group diagram that $S_{\mathcal{U}}$ is represented by an open polygonal path, so $\mathcal{U}$ does not bound a disc in the complement space, but only a surface of higher connectivity, as it must, since K is knotted.

We shall now consider the *Poincaré space of the trefoil knot* (see p.116): a curve $\mathcal{R}$ of the ring surface which cuts the curves $\mathcal{U}$ and C_1 once is represented by the substitution $S_{\mathcal{R}} = C_1 \, C_2 \, C_3 \, C_2^{-1} \, C_2^{-1}$. If we add a disc along $\mathcal{R}$ to the complement space we obtain *Poincaré* space Φ bounded by a spherical surface. Since all curves of the complement space are reducible to curves of the ring surface, the group of this space is equal to the group of curves on the ring surface relative to Φ, i.e. it is the following:

generating elements: C_1, C_2, C_3, C_4

relations: 1) $C_1 \, C_4^{-1} \, C_2 =$ 2) $C_2 \, C_4^{-1} \, C_3 =$ 3) $C_3 \, C_4^{-1} \, C_1 = 1$ $\Big\}$ G_Φ

4) $C_1 \, C_2 \, C_3 \, C_2^{-1} \, C_2^{-1} = 1$

We shall construct the group diagram: for this purpose we consider the following dodecahedral Figure 12. The arrows and numbers associated with segments mean that the segments, taken in the direction of the arrows, represent the generating operation with the index corresponding to the number. The figure consists of nine pentagons, one heptagon and one decagon and the boundary polygon. Each

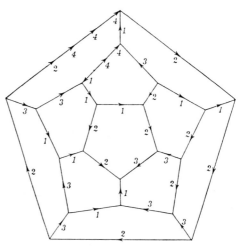

Figure 12

polygon represents, when traversed in a definite sense, either the left hand side of the fourth relation or also this expression transformed by one of the first three relations. We note that the four relations yield the relation

$$5) \quad c_2^5 \, c_4^3 = 1$$

(which corresponds to the boundary polygon). On the other hand, by transformation of the fourth relation with the help of the three others we obtain:

$$6) \quad c_2^5 \, c_4^{-3} = 1$$

from which it follows that

$$7) \quad c_1^{10} = c_2^{10} = c_3^{10} = c_4^6 = 1 \, .$$

Before we use these relations we shall derive something else
in conjunction with the assertion that the relations corresponding
to Figure 12, as well as 6), follow from 4) by means of 1), 2), 3).
For this purpose we construct a part of the group diagram of the
group of the trefoil knot in which the
expression on the left hand side of
4) is represented. This part consists of five strips (see Figure
13). It follows from the relation 4) that the two free
boundaries coincide, and indeed in such a way that the
end points of the polygons
representing expression 4)
coincide, and correspon-
dingly the other points
of both boundaries. It
then follows by direct

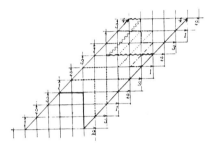

Figure 13

consideration of the figure that the other relations hold as
claimed: the polygons representing them end in corresponding
points of the two boundaries. (In Figure 13 the label of a segment is given on the segment which results from it by horizontal
displacement. Of the segments C_4 only those on the boundary are
represented, and the direction of only one of these is given.
Relation 4) is represented in the figure by a heavily drawn line,
those relations corresponding to polygons of the dodecahedral figure
are represented by dotted and half-dotted lines respectively, while
that of relation 6) is represented by a wavy line.)

Using relation 7) we now obtain the following construction of
the group diagram: we construct a *ring surface* from the five strips
represented in Figure 13, in which we bring the two free boundaries
into coincidence, corresponding to relation 4); the relation $C_4^6 = 1$
further corresponds to any chain of six segments C_4 being closed.
In this way we obtain a ring surface. Each polygon of the
dodecahedral figure is a non-separating closed curve on this ring
surface; we can therefore pass such a ring surface through each of

the polygons. Since three segments meet at every vertex of the
dodecahedral figure, and the segments correspond to three different
vertices, then in the totality of points and strips on the 12 ring
surfaces which meet pairwise along the length of a strip and triply
at a corner we have the desired group diagram. *The group of our
Poincaré space is therefore finite*, it consists of 120 substitutions,
for there are 30 vertices on each ring surface, but each vertex lies
on three ring surfaces, so we obtain the number of substitutions as
$\frac{12 \cdot 30}{3}$ = 120. This number suggests the conjecture that *the group is
isomorphic to the icosahedral group extended by reflection*, a conjecture which is easy to verify. In fact C_1, C_2, C_3 may be regarded
as rotations about three vertices of an icosahedron combined with
reflection in the midpoint, C_4 as the rotation about the midpoint
of the icosahedron combined with reflection in the midpoint.

*The fundamental group of the Poincaré space corresponding to
the trefoil knot is isomorphic to the order 120 group of the icosahedron extended by reflection.*

It is easy to find the groups of *other Poincaré spaces corresponding to the trefoil knot. They are,* with the natural exception
of the trivial case of ordinary space, *all infinite*. The simplest
case is obtained when one adds to the G_K of the trefoil knot the
relation:

$$8) \quad C_1 \ C_2 \ C_3 \ C_1 \ C_2 \ C_3 \ C_2^{-5} = 1 \ .$$

Geometrically, this means that one adds to the complement space an
elementary piece of space along a curve cutting $\tilde{J}$ twice and $\mathcal{U}$
twice. The resulting Poincaré space has the group which is given
by the generating operations with the relations 1), 2), 3), 8).
With the help of the relations we can transform 8) (analogously
to the previous transformation) into

$$9) \quad C_2^{-11} \ C_4^{6} = 1 \ .$$

We see that the group diagram is constructed as follows: We let
the boundaries of a sheet of 11 strips coincide and thus obtain a
cylinder. Further, we take a regular net in the Lobatchevsky plane,
consisting of 11-gons which meet triply at each vertex. We put

cylinders through the interstices of the net, having strips pairwise in common and vertices triply in common. The cylinders are positioned in such a way that the three strips at a corner coincide just as they did in the group diagram of the trefoil knot. It is not difficult to see that this is possible. The vertices and edges on these cylinders give the diagram of the given group, which is therefore infinite. Similar statements are true for the Poincaré groups generated by the curves of the ring which cuts $\widetilde{S}$ n times and $\mathcal{U}$ once. We have the result:

All Poincaré spaces corresponding to the trefoil knot, with the exception of ordinary space and the space studied above have infinite groups, whose diagrams are generated by regular tessellations of the non-euclidean plane in which three (6n-1)-gons meet at each vertex (n > 1).

In order to present the meaning of the group diagram more strongly, we remark that each closed curve on the diagram corresponds to a substitution which is =1, and to a curve on the manifold contractible to a point, i.e. one which bounds a simply connected piece of surface. The above arrangement of the group diagram not only helps to prove that space curves are knotted, or that the spaces are not homeomorphic to ordinary space, but *it also solves the problem of deciding, for a given curve of the complement space or Poincaré space respectively, whether it is contractible to a point or not.*

(b) *Other knots.* Here we shall only deal with the groups of a few especially simple knots, closely connected with the trefoil knot. The first two of the series are shown in the figures 14, 15, the others are obtained by proceeding analogously.

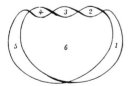

Figure 14

Figure 15

They have $5,7,\ldots,3+2n$ crossings and go over to the trefoil knot by $1,2,\ldots,n$ switches of a crossing point. One easily sees from the rule given in the above paragraph that for $n = 1$ the group is given by

$$\left\{ \begin{array}{l} \text{generating operations:} \quad C_1, C_2, C_3, C_4, C_5, C_6 \\[2ex] \text{relations:} \left\{ C_1\, C_6^{-1}\, C_2 = C_2\, C_6^{-1}\, C_3 = C_3\, C_6^{-1}\, C_4 = C_4\, C_6^{-1}\, C_5 \right. \\[1ex] \qquad\qquad = C_5\, C_6^{-1}\, C_1 = 1 \end{array} \right.$$

The group diagram consists of similar strips as for the trefoil knot (see Figure 16), any five strips meet at a C_6 chain. Similarly for $n > 1$. If we add to these relations, in the case $n = 1$, the new relation

$$C_1\ C_4\ C_2\ C_5\ C_3\ C_2^{-4} = 1$$

then the group represents a Poincaré space corresponding to the knot. One sees that the group diagram is generated by cylinders

Figure 16

made from nine strips, which meet five-fold along the C_6 chains. The cross section of the cylinders corresponds to a regular tessellation of the hyperbolic non-euclidean plane in which five 9-gons meet at each vertex. A similar result holds for $n > 1$. Thus we have:

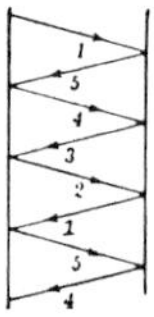

The Poincaré spaces corresponding to the above knots have similar infinite groups, which are generated by regular tessellations of the hyperbolic plane.

Outside of the icosahedral group and the identity we have thus obtained only infinite groups for Poincaré spaces. For manifolds with torsion the only known finite groups are cyclic.

Further investigation of knot groups, in particular as well as general, i.e. the investigation of properties characteristic of all knot groups, will be left to later works.

CHAPTER III Three dimensional manifolds

§1 General (construction of the seam complex and the fundamental group)

Let M_3 be a closed three-dimensional manifold, s_3^1 an arbitrary constituent piece of the same, Π_2^1 the spherical surface bounding s_3^1. Let s_3^2 be a constituent piece of space that has at least *one* disc in common with s_3^1. Using s_3^1 and s_3^2 we construct a 3-cell Σ_3^2 in which we join s_3^1 and s_3^2 together along *one* common disc. The spherical surface Π_2^2 bounding Σ_3^2 consists only of faces which belong to M_3, but a pair of such faces on Π_2^2 may correspond to only a single face in M_3, namely, when s_3^1 and s_3^2 have more than one disc in common. We now attach to Σ_3^2 a third piece of space s_3^3 of M_3, which has at least *one* disc in common with Π_2^2, and in fact we attach along such a face. Σ_3^2 and s_3^3 together constitute Σ_3^3, with boundary Π_2^3. We proceed in this way until we obtain a 3-cell Σ_3 which contains all the constituent pieces of space of M_3, and is bounded by the spherical surface Π_2. Each face of Π_2 also appears in M_3, but a pair of faces of Π_2 correspond to one and the same surface in M_3, since M_3 is closed by hypothesis. We call Π_2 a *"cut surface of M_3"*. M_3 *is therefore homeomorphic to a 3-cell* bounded by a cut surface *when the faces of the latter are identified in the way resulting from the above construction* (first general method of generating M_3). The relationship of the faces of Π_2 to one another also gives their orientation. From the definition of two resp. one-sidedness of multidimensional manifolds we immediately have the result: if two corresponding surfaces of Π_2 always have different orientations in relation to Π_2 given by the above procedure then M_3 is two-sided, otherwise one-sided. So the Π_2 gives us a convenient means of deciding this first question about the character of M_3. We now let corresponding faces of the cut surface Π_2 coincide and obtain a surface complex N_2, which we call the *seam surface* of M_3. This notation indicates that one obtains the original M_3 from a spherical space by stitching together along N_2. In the case of two dimensional manifolds a cut system corresponds to C_2 , by means of which M_2 may be transformed into a simply connected piece of surface. We therefore have the following result:

(1) Each line segment complex and each surface complex of M_3 is homotopic to a line segment resp. surface complex of the seam surface (i.e. it may be continuously deformed into one, allowing self-intersection), because each line segment or surface complex not in the seam surface N_2 lies in the spherical space Σ_3 and thus is homotopic to a complex on the boundary Π_2 and hence to one such on N_2 . Hence if a curve (and also a curve complex) bounds a disc resp. a piece of surface of higher connectivity in M_3 , then a corresponding curve (or corresponding curve complex) of N_2 bounds a disc resp. surface of higher connectivity on N_2 , and conversely. If two curve systems on N_2 are not continuously deformable into each other this is also the case for M_3 ; in particular if a curve on N_2 is not contractible to a point (i.e. bounds no disc of N_2), then it is also not contractible to a point in M_3 . For this reason *we define the fundamental group of N_2 to be the fundamental group of M_3:* $G_{N_2} = G_{M_3}$. The construction of G_{N_2} through representation of bounding circles by a suitably chosen system of curves through a point on N_2 is developed in Chapter I, §2.

Underlying N_2 we have a line segment complex which we shall denote by N_1 . The neighbourhood of this complex, in the case where M_3 is two-sided, is bounded by a two-sided closed surface of genus $p = \mu = \nu_1 - \nu_0 + 1$ (ν_1: number of segments, ν_0: number of vertices of N_1) (see Chapter I, §2). This neighbourhood J_3 consists of a 3-cell with p further pieces of space ("handles") attached. Thus it is representable in our ordinary space. We can call it the *ordinary two-sided multiple ring space*. Likewise, the part of M_3 which remains when J_3 is taken away, the "complement space" A_3

of N_1, will be such an ordinary multiple ring space, as follows immediately from the construction. In the case that M_3 is one-sided, J_3 and A_3 are bounded by a one-sided surface of even characteristic. J_3 and A_3 constitute ordinary one-sided multiple ring spaces. Thus we have a second method of generation:

The general homogeneous closed M_3 is obtained by fusing the surfaces of two ordinary multiple ring spaces.

It follows from this that

Each homogeneous closed M_3 is decomposable into four nonsingular 3-cells.

Each curve of the space is isotopic to a curve on the surface F which separates A_3 and J_3. Then without going any further one can derive the Poincaré theorems on the relation between curves and surface connectivity, which we will not pursue further here. It need only be remarked that these considerations may be immediately extended to any number of dimensions, so that, among other things, it follows that any homogeneous M_n may be decomposed into four n-cells, so there is also a simpler proof of the corresponding Poincaré theorems for an M_n (n > 3).

§2 Ordinary space

One can see immediately that the fundamental group of ordinary space is the identity, since each curve in space bounds a disc (possibly with singularities), so that the same is true of each curve on N_2, from which it follows by Chapter I, §2, that the fundamental group of N_2, and hence M_3, is the identity.

If conversely the fundamental group G_{N_2} of N_2 is the identity then each curve on N_2 is bounding. Under this hypothesis N_2 contains no closed two-sided surface, from which it follows that no open segment of N_2 alone bounds a disc. Then if N_1 contains an open segment we can remove this by pulling together, i.e. if we insert discs in the modified N_1 by suitable circles, then these, taken doubly, again constitute a sphere, and if we identify the two halves of the corresponding 3-cell boundary in the manifold, then we obtain a homeomorph of the given M_3. In this way we can remove all open segments of N_1: the latter then consists of pure circles with *one* common point 0. Now if G_N is the identity, each of these circles bounds a disc, whose only singular lines are the circles through 0. If we could now show that none of these circles had essential boundary singularities (see Chapter I, §3), then it would follow immediately from the lemma that there is a system of non-singular discs which each had one of the circles as boundary and no point in common apart from 0. However, the neighbourhood of such a surface complex in a homogeneous M_3 is a spherical space. Thus we can embed N_1 in a spherical space E_3^i. But, the boundary K_2 of this spherical space lies in a suitably chosen complement space A_3 and in this bounds a 3-cell E_3^α, by the argument of the previous paragraphs. Because each closed surface is bounding in the ordinary multiple ring space, in particular a spherical surface bounds a 3-cell and indeed this holds whether or not we use the assumption that the ring space is two-sided. However the E_3^α and E_3^i, when fused along the bounding spherical surface K, result in the given M_3, and this is consequently homeomorphic to ordinary space. However our assumption of the non-existence of essential boundary singularities is not yet established. Therefore it is necessary to study further the cases in which G_N is the identity.

This famous and influential paper is a memorable introduction to topology, combinatorial group theory and non-
euclidean geometry.   To see Dehn weave these threads into a
coherent pattern is an unforgettable demonstration of the
unity of mathematics.   Of course, there is a historical basis
for this coherence in the work of Klein and Poincaré in the
1880's, and the relation between surface topology and hyperbolic
geometry had already been used with considerable sophistication
by Poincaré [1904], but Dehn's paper is the first, I believe,
to show equal sophistication in the treatment of combinatorial
group theory.   In particular, it is the first in which a purely
combinatorial algorithm is justified by appeal to the hyperbolic
metric.

The result in question is Dehn's first combinatorial
algorithm for the conjugacy (transformation) problem, and his
method of proof is probably its most important part.   The
algorithm itself seems, if anything, _less_ efficient than the
analytic algorithm he wants to supplant, and in any case he
found a much better algorithm later (Dehn [1912b]), with a
purely combinatorial proof.

Suppose that   c, d   are conjugate elements of the fundamental group $\pi_1(\bar{F})$ of $\bar{F}$, so that

$$c = ede^{-1}$$

for some element $e \in \pi_1(\bar{F})$. We then have a quadrilateral
(Fig. 1) in the group diagram

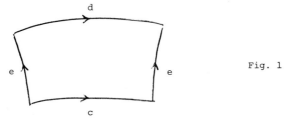

Fig. 1

of $\pi_1(\bar{F})$, the sides of which are edge paths (<u>not</u>, in general,
single edges) whose labels spell out words for c, d, e as
indicated. In general we call a path whose labels spell a word
for $g \in \pi_1(\bar{F})$ a g-<u>path</u>, and a sequence of such paths laid end
to end a g-<u>chain</u>. We know that the endpoints of successive
g-paths in a g-chain lie on a distance curve, hence in the
situation of Fig. 1 we get a c-chain and d-chain running along
distance curves with the same ends R, S on $\partial\tilde{F}$, as in Fig. 2
(Cf. Fig. 1 in the introduction to Dehn's 1910 lectures on
surface topology. The present c-chain corresponds to the
lift chain we called ... $\tilde{c}^{(-1)}\tilde{c}^{(0)}\tilde{c}^{(1)}$ ... there.)

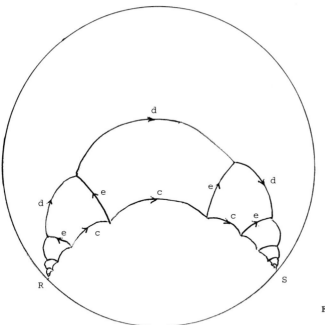

Figure 2

Now if we are given any c, d as words in the generators
of $\pi_1(F)$, and if we want to know whether there is an element e
as in Fig. 1, then hyperbolic geometry suggests that we can
decide this by searching within a bounded hyperbolic distance of
any c-path.   Namely, it is a property of the displacement which
maps a c-chain onto itself that all distance curves through its
ends on $\partial \tilde{F}$ are also mapped onto themselves <u>and</u> that the length
of displacement along a given distance curve increases beyond
all bounds with the distance from the axis.   Therefore, if we
go far enough to either side of the c-chain we shall reach a
distance at which the displacement exceeds the length between

the endpoints of d, beyond which it is useless to search for

a d-path connecting ends of identically labelled paths issuing

from the ends of a c-path as in Fig. 1.   This is the first idea

underlying Dehn's proof.

The second idea is that the <u>word distance</u> between vertices

A, B in the group diagram (i.e. the minimum number of edges in

a path connecting A and B, which is the number of letters in

the word labelling the path) is roughly comparable to the hyper-

bolic distance between A, B.   This is formalised by Dehn's

lemma which states that, for suitably chosen unit of hyperbolic

length,

$$\text{word distance} \leq 4p(p+1) \times \text{hyperbolic distance}$$

where p is the genus of F.   It follows that we can use the

lengths of the words c, d, rather than the hyperbolic metric,

to compute a bound on the (word) distance we need to travel on

either side of the c chain in search of a d-chain tied to the

c-chain by e-paths as in Figure 2.   The search is conducted

by emitting identically labelled paths, e', from the ends of a

c-path.   If the e'-paths grow together, one edge at a time,

then they will hit the d-chain simultaneously, at which moment

there will be a d'-path connecting their free ends, where
d' is a cyclic permutation of d.   We can therefore observe
contact with the d-chain by checking at each stage whether
a d'-path joins the free ends of the e'-paths.   If contact
does not occur within the computed distance, then the required
d-chain does not exist.

The computational catch in this algorithm is that the
e'-paths are found essentially by trying all paths out of the
ends of a c-path until one reaches a vertex whose displacement
under the motion of the c-chain is sufficiently large.   Since
the number of such paths grows exponentially with the length
of the word c, the algorithm is not computationally feasible
as it stands.   Nevertheless, the proof captures a very
interesting relation between analytic and combinatorial facts —
the rough comparability of the word metric and the hyperbolic
metric - which might well have been lost had Dehn thought of
his purely combinatorial proof sooner.   This relationship
remains important and has been taken up again recently, e.g.
by Floyd [1980], Cannon [1984] and Gromov [1981].

REFERENCES

J. Cannon [1984] : The combinatorial structure of cocompact
        discrete hyperbolic groups. Geom. Dedicata, 123-148.

M. Dehn [1912b] : Transformation der Kurven auf zweiseitigen

    Flächen. <u>Math. Ann.</u> 72, 413-421.

W. Floyd [1980] : Group completions and limit sets of Kleinian

    groups. <u>Inv. Math.</u>, 57, 205-218.

M. Gromov [1981] : Hyperbolic manifolds, groups and actions.

    <u>Riemann Surfaces and Related Topics</u> (Eds. I. Kra & B. Maskit),

    Princeton Univ. Press, 183-213.

H. Poincaré [1904] : Cinquième complément à l'analysis situs.

    <u>Rend. circ. mat. Palermo</u> 18, 45-110.

ON INFINITE DISCONTINUOUS GROUPS

INTRODUCTION

The three fundamental problems for infinite discontinuous groups

The general discontinuous group is given by n generators and
m relations between them

$$R_1 (S_{i_1} \ldots) = 1$$

$$\cdot \quad \cdot \quad \cdot \quad \cdot \quad \cdot \quad \cdot \quad \cdot$$

$$R_m (S_{i_m} \ldots) = 1,$$

as first defined by Dyck (Math. Ann., 20 and 22).   The results of
those works, however, relate essentially to finite groups.   The general
theory of groups defined in this way at present appears very undeveloped
in the infinite case.   Here there are above all three fundamental
problems whose solution is very difficult and which will not be possible
without a penetrating study of the subject.

1. **The identity problem [word problem, trans.]** : An element of
the group is given as a product of generators.   One is required to give
a method whereby it may be decided in a finite number of steps whether
this element is the identity or not.

2. **The transformation problem [conjugacy problem, trans.].**
Any two elements S and T of the group are given.   A method is

sought for deciding the question whether S and T can be transformed into each other, i.e. whether there is an element U of the
group satisfying the relation

$$S = UTU^{-1}.$$

3. The isomorphism problem : Given two groups, one is to decide
whether they are isomorphic or not (and further, whether a given
correspondence between the generators of one group and elements of
the other group is an isomorphism or not).

These three problems have very different degrees of difficulty.
Problem 1 is a special case of problem 2.   The first obviously
concerns the character of one element of a group, the second the
relation between two elements of one group, the third the relation
between the elements of two groups.   It may be presumed that the
solution of these three problems would constitute a natural foundation
for the methodical presentation of the theory of infinite groups.
However, one is already led to them by necessity with work in topology.
Each knotted space curve, in order to be completely understood topologically, demands the solution of the three above problems in a
special case : each curve K is unknotted iff G_K is an abelian group,
which leads to an identity problem[*].   Any other curve in space in
relation to K corresponds to a definite element of G_K.   Two
space curves are then transformable into each other without cutting K
iff the corresponding elements of G_K are transformable into each

---

[*] See Math. Ann. 69.

other.   Finally, the question whether a given space curve K' can

be transformed into K without self-intersection requires a solution

of the third problem for $G_{K'}$ and G_K.

The appearance in topology also dispels certain objections which

might be raised against the restriction to infinite groups which can be

defined in the above manner.   In fact, such objections can arise when

one considers that in general, each infinite group has subgroups which

cannot be generated by a finite number of generators and relations between

them.   [Actually, there are infinitely many exceptions, such as $\mathbb{Z}^n$ - trans.]

Consider, say, the group which is given by two generators S_1 and S_2

without defining relations, and collect all those elements which result

from S_1 by composition and transformation in the group, e.g.

S_1, S_1^n, $S_2 S_1 S_2^{-1} S_1$, etc.   These elements constitute a group, and indeed

a normal subgroup of the given group, whose elements cannot be generated

by any finite number of them.   This group therefore does not come under

the heading of the groups defined above.

Each of the three problems for finite groups is solvable without

further effort and therefore meaningless.   If one confines oneself

to analytic groups, i.e. groups of transformations in an n-dimensional

real or complex number manifold, in the case where the generators are

given directly by certain transformations, the first problem is solved

in most cases without further consideration.   The first problem

may likewise be solved by constructing the group diagram

corresponding to the given group, i.e. the infinite line segment complex which was introduced in the work in Math. Ann. 69 and also made the basis of that work.

If the group is given as a proper discontinuous group of motions in euclidean or non-euclidean space, then the second problem is settled by construction of the fundamental domain associated with the group. Likewise we obtain the solution of the problem in the general case by canonical representation of the self-motions of the group diagram.

The analytic representation is less useful for the solution of the third problem. Topological investigation of the group diagram appears to be of essential importance here.

In the present work the three problems are settled for the case where each generator appears only twice altogether in the defining relations (Chap. II). In order to attain this, we must begin with an exact study of the simplest class of such groups, namely the fundamental groups of closed surfaces (Chap. I). Finally (Chap. III), a few theorems on the most general groups are given, and a few examples of higher groups are treated. Among other things, the transformation problem for some knot groups is settled here, the ones for which the identity problem has already been solved in Math. Ann. 69 by setting up the group diagram.

Chapter I

The fundamental groups of closed surfaces

Strictly speaking, one cannot say that pure group theory is of great service to topology in the case of fundamental groups of closed surfaces. On the contrary, the solution of the topological problem bears fruit for group theory, for in certain simple cases that solution immediately settles the three fundamental problems of group theory, and indicates a way that leads to the same goal in many other cases.

Thus we must first explain topological matters, and in fact these concern the continuous transformations of closed curves on surfaces. Here we have to be somewhat more detailed than would be necessary had this theory, whose important results are known*, at least for two-sided surfaces, been carried out anywhere else.

§1. Two-sided surfaces

1. By cutting along 2p circuits which all go through a point and fall into successive pairs of intersecting curves $a_1, b_1, a_2, b_2, \ldots,$ $a_p, b_p,$ the closed two-sided surface of genus p is transformed into a 4p-gon, whose edges are identified with each other in pairs.   If one denotes by a_i resp. a_i^{-1} etc. the curve a_i with its two directions, then the edges of the 4p-gon give the sequence

*) Poincaré, Pal. Rend. 1905 p. 14 ff.   The great work of C. Jordan (J. d. Math. 1866), which is entirely devoted to this problem, leads to an absolutely incorrect result. One also sees clearly here the substantial simplification brought about by the application of non-euclidean geometry.

$$a_1, \ b_1, \ a_1^{-1}, \ b_1^{-1} \ ; \ a_2, \dots, a_p^{-1}, \ b_p^{-1}$$

when one traverses the boundary in a definite direction.   Each closed curve on the surface becomes, in the 4p-gon, either a closed curve or curve pieces $\delta_1, \ \delta_2, \ \dots$ each bounded by two boundary points of the 4p-gon.   The final point of δ_i and the initial point of δ_{i+1} are homologous points on corresponding edges or else both vertices of the 4p-gon.

2. As one easily sees, one can construct an infinite net of the following character without making any metric assumptions, i.e. purely topologically : 4p line segments emanate from each vertex, each cell is a 4p-gon.   The edges of the net can be denoted in such a way that in each cell the series of edges follow each other in the same order

$$a_1, \ b_1, \ a_1^{-1}, \ b_1^{-1} \ ; \ a_2, \dots, a_p^{-1}, \ b_p^{-1}.$$

Then at each vertex the line segments taken in the same order are

$$b_p, \ a_p^{-1}, \ b_p^{-1}, \ a_p, \dots, b_1^{-1}, \ a_1.$$

Of course a segment which is denoted in one sense by a_i will be denoted by a_i^{-1} when traversed in the opposite sense.   It is suitable for what follows to take the net for $p = 1$ to be a net of squares in the euclidean plane, and for $p > 1$ to take the net of regular 4p-gons in the hyperbolic plane.

3. These nets can be viewed more familiarly as mappings of the

two-sided surface covered infinitely often, where any two covering

sheets are connected along one circuit cut.   Points of different

sheets which correspond to the same point of the surface will

appropriately determine homologous points, in the metric sense,

of different cells of the net.   A connected curve on the surface

determines a likewise connected curve of the net, which enters new

cells as often as the original curve cuts circuits on the surface.

The initial and final point of the image of a closed curve are homologous

points of the cells concerned.   Each closed curve of the surface

determines, by its image in the net, a   motion   of the net into

itself, in which each net segment goes into one with the same

denotation, and the initial point of the image goes into the final point.

Here a definite sense is associated with the curve on the surface.

The net   motion   which carries the final point to the initial point

corresponds to the curve traversed in the opposite direction.

4. If the final point coincides with the initial point then the

corresponding net   motion   is the identity ; for when the denotation

and one point of the net are fixed there is no motion of the

net onto itself except the identity ; correspondingly, given that a

point   a   shall go to a homologous point   a', the motion is

uniquely determined.

If the image of a curve on the surface is a closed net curve,

then the surface curve is contractible to a point; for as the

mapping shows, the surface curve bounds a simply connected piece

which in general will cover a piece of the surface with a certain multiplicity.   Conversely, the image of a surface curve contractible to
a point is a closed curve of the net.   For such a curve may be
continuously ("isotopically") transformed on the surface into an
arbitrary curve contractible to a point, the image of which may be
chosen to lie in a single cell and hence must be a closed curve.
But since continuous transformations on the surface correspond to
continuous transformations on the net, the image of the original
curve must also be closed.   Since we can decide whether a curve on
the net is closed or not without metric help, the question whether
a given curve on the surface is reducible to a point is directly
answered in this way by pure topology.   It is assumed that the surface
curve is given by the sequence of its intersections with the circuit
cuts, and correspondingly the image curve on the net is given in a
similar way by successive points on the edges of the net.

5. The image of a surface curve not contractible to a point
has infinitely many forms depending on the choice of initial point.
If one chooses a certain initial point for the surface curve, but
different (homologous) points as images, then the surface curve corresponds to different net motions ,   but ones which transform
into each other* by   motions   of the net.   If we choose another
point as the initial point of the surface curve, but retain the
original image of the original final point, then the associated motion
of net remains unaltered.   If one fixes the initial point of the

---

*
 I.e. are conjugate (Translator's note).

surface curve and continuously deforms the latter on the surface, then the associated initial and final point of the net curve are retained. The associated net motion is therefore likewise unaltered. Conversely, all curves bounded by two fixed net points are images of homotopic surface curves. Now since any continuous transformation of a surface curve can be assembled from transformations in which a point of the curve remains fixed, we have the result :

Two surface curves are continuously transformable into each other on the surface iff any pair of net motions associated with the surface curves are transformable into each other.

Deciding the question whether two surface curves are transformable into each other or not is therefore equivalent to the solution of a problem concerning the euclidean resp. non-euclidean metric. Without going further we cannot give a purely topological method for settling the general problem. This is only possible in the case $p = 1$, i.e. in the case where the net lies in the euclidean plane. Here all motions result from parallel translations in the directions of the a and b sides of the net respectively. The motion group of the net is abelian, and two motions are transformable into each other if they are composed of equal numbers of a- and b-translations. This leads in an obvious way to a simple topological resolution of the question.

In the case $p > 1$ we have to deal with motions (without reflection) in the hyperbolic plane. A single such motion is equivalent to a rotation about a real or ideal midpoint. In our case it must be ideal, since there is no motion   different from the identity with a fixed real point, as already remarked. Thus each motion under consideration is a translation along a real axis. Then, in order for two such motions   to be transformable into each other it is necessary and sufficient that the corresponding displacement segments along the axes be composed of homologous pieces in the same way. Thus if we represent the displacement segments by segments which each lie in a single cell, then the motions   are transformable into each other iff these unions of segments are identical. In this way our problem is reduced to an algebraic problem which is to be settled in a finite number of steps: in fact, 1. the coordinates of the net points are roots of algebraic equations; 2. the net motions   are representable as linear transformations with coordinates which are known algebraic functions of the coordinates of the net points; 3.   therefore, the coefficients of the equation of the displacement axis are known algebraic functions of those quantities, and the same holds for the coordinates of the intersection point of the axis with the net sides; 4. the solution of our problem then comes down to deciding whether certain roots of given algebraic equations are identical with each other, which can be settled in a finite number of steps by known theorems.

While the foregoing general remarks are already bringing the conviction that our problem may be settled in a finite number of steps, it is still unsatisfactory that the method implied is too

intricate to be of practical value in particular cases.   In this
connection the following practical method, based on the foregoing
considerations, is of significance.   Its presentation imposes on
us the pleasant duty of exploring more deeply the geometric
properties of polygonal nets.

6. Relative to a vertex A, all vertices of the net [*] can be
divided into classes as follows: the first class contains A, the
second class contains all vertices of the 4p net edges emanating
from A, the third the vertices of the edges emanating from points
of the second class, etc., the points of the mth class being the
vertices of edges emanating from points of the $(m-1)^{th}$ class, as
long as these do not belong to preceding classes.   A point of the
m^{th} class, B, say, can be connected to A by an (m-1)-step polygon
edge   path, but by no shorter path.   We call (m-1)s, where s
is the length of the net edge, the net separation ε_N of A and B :

$$\varepsilon_N(A,B) = (m-1)s.$$

Further, $\varepsilon(A,B)$ shall denote the straight line separation
between A and B.

Lemma : $\varepsilon_N(A,B) \leqslant 4p(p+1)\varepsilon(A,B).$

We obtain the proof in two stages.

I. The distance between two points on the boundary of a net

[*] In this section we fix p > 1.

polygon which lie on two non-adjacent edges is greater than $\frac{s}{2p}$.

a) We connect the midpoints of two adjacent edges by a circular arc with centre at the common vertex. Since the lines connecting the midpoint of the polygon with the midpoints of the edges are perpendicular to the edges and lie entirely inside the polygon, the circular arc also lies entirely inside the polygon. The distance of a vertex of the polygon from a point on a non-adjacent edge is therefore $> \frac{s}{2} > \frac{s}{2p}$.

b) If e and f (see Fig. 1) are two non-adjacent polygon edges, then there is always a line segment within the polygon perpendicular to both edges and hence equal to the shortest distance between them.

Fig. 1.

Then if we connect the ends of e and f by two non-intersecting chords, a quadrilateral with four acute angles results of which any two on a chord are equal, since e and f are equal chords in a circle. (The angles are smaller than the polygon angle, therefore acute, since we assume this of the polygons.) The feet of the common perpendicular lie at the midpoints of the edges when these are opposite each other, otherwise nearer to the endpoints of the shorter chord, since the perpendicular bisectors of the edges cut the longer chord.

If AB and AC are two edges of the polygon, M_1 and M_2

their midpoints and M the midpoint of the polygon (see Fig. 2)

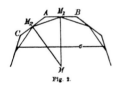

Fig. 2.

then it follows from the congruence of the triangles M_2AM_1 and M_2MM_1 that $\angle AM_1M_2 = M_2M_1M = \frac{\pi}{4}$ and that $MM_1 = AM_1 = \frac{s}{2}$. Now M_1, M_2 etc. describe a regular $4p$-gon inscribed in the circle with radius MM_1, so that

$2pM_1M_2$ is greater than $2MM_2$, the diameter of the circle, and thus $M_1M_2 > \frac{s}{2p}$.

Now if σ is a segment of shortest length between two edges which are separated by at least two edges, then σ cuts off a piece of the circle with centre M, radius MM_1, which is not greater than a half circle and in which one chord M_1M_2 is completely contained (Fig. 2) ; consequently $\sigma > M_1M_2 > \frac{s}{2p}$. Finally let DE be a segment of shortest length between two edges which are separated

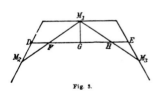

Fig. 3.

only by a single edge (see Fig. 3), and connect M_1 and M_2, M_1 and M, M_1 and M_3 by lines which cut DE in F, G, H respectively. It follows from the congruence of the triangles M_1GF and M_2DF (the three angles are respectively equal) that $DF = FG$, $M_2F = M_1F$. Since now, as was shown earlier, $\angle M_2M_1M_3 = \frac{\pi}{2}$, it follows that

$$DE = 2FH > 2FM_1 = M_1M_2 > \frac{s}{2p} .$$

With this the first objective is reached.

II. A   segment   connecting two vertices of the net can
be divided into pieces which by the foregoing are $> \frac{s}{2p}$:
a) diagonals of a polygon, b) lines connecting a vertex and a point
on a non-adjacent edge of the same polygon, c) lines connecting
two points of the same polygon which are on two non-adjacent edges,
d) segments of the form XY shown in Fig. 4 :   the number of

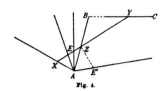

Fig. 4.

edges emanating from A met by
XY is arbitrary.   YC belongs
to the same polygon as AB, but need
not be adjacent to AB at B.
However, Y can also coincide with
B and Z.   ZE' is the mirror

image of ZE with respect to AB.   It follows that:

$$XY \geqslant EY = YZ + ZE' > YE' > \frac{s}{2p} \; .$$

Thus also $XY > \frac{s}{2p}$.

In the cases a), b) and c) one can connect the endpoints of a
subsegment by a path of successive polygon edges,
the length of which is $\leqslant 2ps$, in case d) by a path whose length
is $\leqslant (2p+2)s$.   The ratio $\frac{\varepsilon_N}{\varepsilon}$ is therefore $\leqslant \dfrac{(2p+2)s}{\frac{s}{2p}}$ for each
subsegment, and hence for the whole segment

$$\varepsilon_N \leqslant 4p(p+1)\varepsilon, \quad \text{as claimed.}^*$$

---

*
In the original, Dehn erroneously reverses the last two inequalities.
(Translator's note)

A motion yields a mapping of net vertices, and is determined by the correspondence between two vertices. If X_0 goes to X_1 under a motion then we shall denote the point which goes into X_0 under the same motion by X_{-1}, the point into which X_1 goes by X_2, etc. If now A_0, A_1 and B_0, B_1 are any two vertex pairs of the net and it is to be decided whether the motion $A_0 \mapsto A_1$ can be transformed into the motion $B_0 \mapsto B_1$, then one connects B_0 with B_1 by an edge path $B_0 \alpha_0 \beta_0 \gamma_0 \cdots B_1$. Further, let the points which correspond to $\alpha_0, \beta_0, \gamma_0, \ldots$ under the motion be $\alpha_1, \beta_1, \gamma_1, \ldots$ (i.e. $B_0 \alpha_0$, $B_0 \beta_0, \ldots$ go into $B_1 \alpha_1$, $B_1 \beta_1, \ldots$). Let the greatest of the net distances $\varepsilon_N(A_0, A_1)$, $\varepsilon_N(B_0, B_1)$, $\varepsilon_N(\alpha_0, \alpha_1)$, $\varepsilon_N(\beta_0, \beta_1), \ldots$ be m. By suitable trials (in which one successively investigates the points of the 2nd, $3^{rd}, \ldots$ classes of vertices with respect to A_0 and A_1 respectively) one obtains on both sides of the edge path $\ldots A_{-1} A_0 A_1 \ldots$ pairs of corresponding points X_0, X_1 and Y_0, Y_1 whose net separations are greater than $4p(p+1)m.$*

One connects A_0 and A_1 with X_0, Y_0 and X_1, Y_1 respectively by edge paths

$$A_0 \xi_0 \xi_0' \xi_0'' \ldots X_0 \quad \text{resp.} \quad A_1 \xi_1 \xi_1' \xi_1'' \ldots X_1$$

and $\quad A_0 \eta_0 \eta_0' \eta_0'' \ldots Y_0 \quad \text{resp.} \quad A_1 \eta_1 \eta_1' \eta_1'' \ldots Y_1.$

Through the vertices $\xi_0, \xi_0', \ldots$ and $\eta_0, \eta_0', \ldots$ one lays all edge paths labelled the same as $B_0 B_1$, $\alpha_0 \alpha_1$, $\beta_0 \beta_1 \ldots$ <u>Then</u> $B_0 \mapsto B_1$ <u>is transformable into</u> $A_0 \mapsto A_1$ <u>iff there is such a</u>

---

*Here Dehn uses the fact that the distance a point is displaced by a hyperbolic translation increases indefinitely with its distance from the axis (Translator's note).

<u>path beginning in</u> $\xi_0^{(i)}$ <u>or</u> $\eta_0^{(i)}$ <u>which ends in</u> $\xi_1^{(i)}$ <u>or</u> $\eta_1^{(i)}$
<u>respectively</u>.

It is clear that this condition is sufficient because any two
paths with the same labels can be carried into each other by a motion.
The necessity of the condition results as follows: by our lemma the
straight line distances $\varepsilon(X_0,X_1)$ and $\varepsilon(Y_0,Y_1)$ are $> m$, and thus
greater than $\varepsilon(A_0,A_1)$; it follows that outside the region between
the circles $\ldots X_{-1},X_0,X_1,\ldots$ and $\ldots Y_{-1},Y_0,Y_1,\ldots$ the separation
between corresponding points is always greater than $\varepsilon(X_0,X_1)$ resp.
$\varepsilon(Y_0,Y_1)$, so their net separation is also greater than the net
separations $\varepsilon_N(B_0,B_1)$, $\varepsilon_N(\alpha_0,\alpha_1)$, $\varepsilon_N(\beta_0,\beta_1),\ldots$. Then if there is
a motion, carrying $B_0 B_1$ into a point pair $B_0'B_1'$ which consists of
two points which correspond under the motion $A_0 \mapsto A_1$, then α_0 and
α_1, β_0 and $\beta_1,\ldots$ would go into pairs of points α_0',α_1' ;
β_0',β_1' ; $\ldots$ which correspond according to $A_0 \mapsto A_1$, and so the whole
edge path $B_0'\alpha_0'\beta_0' \ldots B_I'$ must lie between the circles $\ldots X_{-1}X_0X_1 \ldots$
and $\ldots Y_{-1}Y_0Y_1 \ldots$ and hence meet the edge path $X_0 \ldots A_0 \ldots Y_0$
in at least one point, from which the necessity of our condition
follows immediately.

§2. One-sided surfaces

If one cuts a closed one-sided surface of characteristic k

(= the maximal number of closed curves on the surface which are jointly not bounding) along k suitably chosen circuits running through a point, $a_1, a_2, \ldots, a_k$, then one obtains a simply connected piece of surface whose boundary, when traversed in a certain sense, consists of the segments $a_1, a_1, a_2, a_2, \ldots, a_k, a_k$. If, in the case $k = 1$, we construct a sphere covered by two 2-gons ; in the case $k = 2$ a square net in the euclidean plane ; in the case $k > 2$ a net of regular $2k$-gons in the hyperbolic plane, where $2k$ of them meet at each vertex, then we can in each case denote the edges of the net in such a way that the boundary of each cell, traversed in a suitable sense, consists of the above $a_1, a_1, \ldots, a_k, a_k$. Then just as in the case of two-sided surfaces we can map each closed curve to a connected curve of the net-bearing plane (resp. sphere), with homologous endpoints. Corresponding to surface curves which are continuously transformable into each other we have motions of the net which are transformable into each other. In the case $k > 2$ these motions are rotations about an ideal midpoint with or without reflection. In fact, with any motion different from the identity no real point remains at rest. Two motions are then transformable into each other iff they are both translations with reflection or both translations without reflection, and further, the representatives of the displacement segments along the axes of both translations are identical in a net cell. The same considerations hold here as those presented for two-sided surfaces in paragraphs 4, 5, 6. These lead us to a method of

deciding the question whether two net motions are transformable into each other in the case $k > 2$, or, what comes to
the same thing, whether two surface curves are continuously transformable into each other.

In the case $k = 1$ we have only one motion different
from the identity, namely rotation of the sphere through 180°,
and correspondingly only one kind of surface curve not contractible
to a point.

The case $k = 2$ still remains. The surface is developed
on a square net in the euclidean plane. The edges of a cell form
the series a, a, b, b. The motions of the net into itself
are either parallel displacements or reflections combined with a
parallel displacement along the axis of reflection. The axis of
reflection connects either the midpoints of edges denoted a, or
else the midpoints of edges denoted b. It is then easy to decide
whether two motions are transformable into each other: two
motions transformable into each other are either parallel
displacements of equal magnitude and direction, or reflections in
homologous axes combined with equal parallel displacements along
the axis of reflection.

§3. Analytic expression for surface curves

Each motion of the net into itself can be generated by

the correspondence between two net vertices. Thus each surface
curve is transformable into one whose net image consists purely
of net edges. Such a curve, or the corresponding net motion,
will be denoted by the names of the net edges taken in the order
corresponding to traversal of the curve from initial to final point
If A is any expression denoting a path of net edges , and B
any other such expression, then

$$B^{-1}AB = C$$

is an expression denoting the most general path transformable into
A. The net motions resp. surface curves corresponding to
the edge paths A and C are transformable into each other
and conversely : if A and C are two edge paths which
correspond to net motions transformable into each other,
then one can always find an edge path B such that the above
relation holds between A, B, C. These remarks hold for all
developments of surface complexes on nets. If we apply this
specially to the two-sided and one-sided ring surface, it follows
that in the first case all paths of net edges are transformable
into an expression

$$a^n b^m$$

where n and m are integers and a^n naturally denotes the
edge path consisting of n segments a. Two such expressions

are then transformable into each other iff their n and m values coincide. The situation with the one-sided ring is somewhat more complicated. We note for future use that in any expression we can delete a sequence aabb, since the corresponding four segments constitute a closed path. We can anywhere replace b by $a^{-1}ab$ and b^{-1} by $(ab)^{-1}a$ (where $(ab)^{-1}$ denotes the edge path ab traversed in the opposite sense). ab will be denoted c for simplicity. We then have:

$$ac = c^{-1}a.$$

Then in each edge path which contains only c and a we can collect all c's and a's together and thus obtain (without transformation) an expression of the form

$$a^n c^m \quad \text{or} \quad c^m a^n$$

— expressions convertible into each other by transformation. If n is even this expression represents a parallel displacement, and it easily follows that no transformation can alter the numbers m or n. So for two such expressions with even n to be transformable into each other, their numbers m and n must coincide. If, however, n is odd and $m = m_1 + m_2$ then by the above

$$c^m a^n = c^{m_1 + m_2} a^n = c^{m_1} a^n c^{-m_2}.$$

So by transformation, $c^{m_1 + m_2} a^n$ goes into $c^{m_1 - m_2} a^n$. Thus any expression with an odd exponent for a goes into one of the forms

$$c^n \quad \text{or} \quad ca^n$$

by transformation.   Summing up we have : two expressions are
transformable into each other iff: 1. the algebraic sums of the
exponents of a are equal, and 2. a) if the sum is even, then the
algebraic sums of the c exponents are also equal, b) if the sum is
odd, the algebraic sums of the c exponents are either both even
or both odd.   With this, the theory of transformations of curves
on the one-sided ring is completely settled.

§4. Amphidrome surface curves

On one-sided surfaces with $k = 1$ or 2 there are curves
not contractible to a point which are transformable into themselves
reversed.   Indeed, for $k = 1$ there is only a single type of
surface curve not contractible to a point.   The curves, twice
traversed, bound a simply connected piece of surface, and thus such
a curve is transformable into itself reversed, a point of the curve
remaining fixed in the process.   For $k = 2$ the curves expressed
by ab, ba respectively in the above notation are transformable into
each other.   However,

$$abba = 1$$

so
$$ab = (ba)^{-1}$$

which says that the curve corresponding to the expression ab is
transformable into the curve corresponding to the expression $(ba)^{-1}$,
continuously and with a point fixed. Consequently the curves
corresponding to the expressions ba and $(ba)^{-1}$, which are the
same except for sense, are transformable into each other. We shall
call such curves amphidrome and show that there are no amphidrome
curves on other closed surfaces.

We first consider the one-sided surfaces with $k > 2$. The
corresponding net motions are rotations about an ideal midpoint
in the hyperbolic plane with or without reflection in the real axis.
Since no real point can remain at rest under a net motion,
a pure reflection without simultaneous rotation is not possible.
No line of the plane apart from the axis goes into itself under the
motion. Thus in order that a motion other than the identity go into
the inverse motion as a result of a second motion, the second motion
must have the same axis as the first. But if this is the case,
the two motions commute. It follows from this that the first motion
performed twice in succession yields the identity. But this is
impossible, since each motion is a rotation about an ideal centre.
The same conclusion also holds for two-sided surfaces when $p > 1$.
If $p = 1$ all motions commute with each other and in fact are parallel
displacements in the euclidean plane. No motion performed twice in
succession can result in the identity, and consequently no motion
can be transformed into its inverse. Amphidrome surface curves

correspond, however, to net motions which are transformable into their own inverses.   Thus we have the theorem:

Amphidrome curves exist only on one-sided surfaces with k = 1 and 2.

For k = 2 it is easy to ascertain all amphidrome curve types.   They correspond to parallel displacements of the net perpendicular to the axis of reflection and hence correspond to expressions of the form $(ab)^m$ or curves transformable into such.

§5.   Introduction of the fundamental group

The motions of the net, on which we have developed a closed surface as above,   constitute a group.   It is clear that we can take as generators of this group the   2p   resp.   k motions corresponding to the   2p   resp. k   closed surface curves along which the surface is cut in order to be developed.   If we denote each motion by the name of the corresponding curve then according to the above notation we have as generators of the group of motions $a_1, b_1, \ldots, a_p, b_p$ resp. $a_1, \ldots, a_k$, between which there is a single relation, namely

$$a_1 b_1 a_1^{-1} b_1^{-1} \ldots a_p b_p a_p^{-1} b_p^{-1} = 1$$

resp.

$$a_1^2 \ldots a_k^2 = 1.$$

In fact each relation between motions of the net may be derived from the defining relation, because the closed curve corresponding to the given relation bounds a region consisting of cells, and the defining relation says that a cell is bounded by a closed curve.

The net is the group diagram[*] of the group of motions of the net into itself. This group is called the fundamental group of the corresponding surface. Each closed surface curve running through a fixed point of the surface corresponds to an element of the group, two surface curves transformable into each other correspond to two group elements transformable into each other, in particular surface curves contractible to a point correspond to the identity. If A and B are surface curves corresponding to elements a and b of the fundamental group, then the element ab corresponds to the surface curve which consists of A followed by B. The identity and transformation problems for these fundamental groups are completely solved by the considerations of §1-3.

The fundamental groups so obtained are isomorphic, as is easily seen, to groups with the same number of generators but a different relation between them. E.g. the group

$$\begin{cases} \text{generators} : & a_1, a_2 \; ; \\ \text{relation} : & a_1 a_2 a_1^{-1} a_2 = 1 \end{cases}$$

is isomorphic to a fundamental group of the above form

$$\begin{cases} \text{generators} : & a, b \\ \text{relation} : & a^2 b^2 = 1. \end{cases}$$

[*] see Math. Ann. 69 (Dehn's "On the topology of 3-dimensional space").

We shall present a condition which is sufficient to ensure that a group with one defining relation is isomorphic to the fundamental group of a closed surface. We shall see later that this condition is also necessary[*]. If we construct the group diagram of the group given by the relation, in which we allow polygons corresponding to the relation and all its cyclic permutations to emanate from each point, then the given group is isomorphic to the fundamental group of a closed surface when these polygons, joined along like labelled sides emanating from the initial point, close to a single sheet. In fact, in this case each generator must appear twice in the relation. A closed surface results when I bend the piece of surface bounded by the polygon corresponding to the relation and identify the sides with the same denotation. Each side becomes a closed surface curve, and it is easy to see that the fundamental group of this surface is the group given, so our assertion is proved.

§6. Topological properties of the group diagrams
of fundamental groups of closed surfaces

We shall now derive a theorem on the group diagram of the most general group isomorphic to the fundamental group of a closed surface. The diagram of such a group in "canonical" form is a net in the euclidean or non-euclidean plane. It is clear that we

---

[*] when each generator appears at most twice in the relation.

can connect any two net vertices[*] by a net edge path which has no point

in common with the vertex that represents the identity. This

carries over in a certain form to the most general representation

of this group ; let $S_1, S_2, \ldots, S_n$ be the generators of the group

in canonical form, $T_1, T_2, \ldots, T_m$ the generators of the general

representation, then by hypothesis the S_i may be expressed in

terms of the T_i. Let h be the greatest number of T_i which

appear in any of these expressions, then in the general represen-

tation we consider the totality of all those net vertices whose "net

separation" (see p.143) from the identity is not greater than h.

We also consider the image B of this collection in the canonical

net. All vertices of the net with the exception of a finite number

may be connected by edge paths having no point in common with the

vertices of B. In the general representation all these elements

are vertices which may be connected by edge paths having no

vertices in common with the identity. In fact : two of these vertices

may be connected, by the above, by an edge path consisting of

terms which are expressions for the S_i in terms of the T_i and

which connect points which have a separation from the identity greater

than h. Consequently, none of these terms touches the identity,

so we have the theorem:

In each representation of the fundamental group of a closed surface

one can connect any two elements, with a finite number of exceptions,

by an edge path which does not contain the identity.

[*] different from the identity (Translator's note).

It follows from this by an easy generalization that:

Given a set* of elements in any representation of the fundamental group of a closed surface, there are only a finite number
of elements which one cannot connect by   edge   paths having no
point in common with the given set.

Further:

In each representation of the fundamental group of a closed
surface, each closed   edge   path satisfying a certain inequality
has the property that a ring-shaped band surrounding it divides all other
elements into two classes, inner and outer.   Any two outer or
any two inner elements can be connected by an edge   path which
has no elements in common with the given closed path.   Each
edge   path which connects an inner and an outer element has at
least one element in common with the closed path.

Thus we see that in a certain sense the simple connection of
the elements in the canonical net is preserved in the general
representation.   Similar considerations evidently apply to any kind
of group : certain connectivity properties of the group diagram
appear as "invariants" of the group, independent of the representation.

To conclude our examination of the fundamental group we prove
the theorem:

Two fundamental groups of closed surfaces are isomorphic iff
they correspond to homeomorphic surfaces.

---

*Of course, the set has to be finite (Translator's note).

Tietze[*] has proved the following theorem on infinite groups:
If one allows exchange of terms in the fundamental relations, then
in the resulting relations between generators some may be expressed in
terms of an arbitrary r of the remainder.   This number r is
the same for two isomorphic groups — for a surface of genus p,
r is equal to 2p, for a one-sided surface of characteristic k
r is equal to k-1.   If our theorem were incorrect, then on account
of the Tietze theorem the fundamental group of a two-sided surface
of genus p would be isomorphic to the fundamental group of a
one-sided surface with characteristic 2p+1.   Now Tietze has
also proved that two surface complexes which have isomorphic groups
have the same torsion numbers.   On the other hand, two-sided
surfaces always have torsion numbers equal to unity.   However, each
one-sided surface has curves which first bound when traversed twice,
so such a surface has a torsion number equal to 2, from which the
desired result follows.

Chapter II

Groups in which each generator appears at most
twice in the defining relations

Our task is to use the investigations of the preceding chapter
to bring the groups referred to in the title into a canonical form

---

[*] Wien. Ber. 1907. (Also, Monatsh. Math. Phys. 19 (1908), 1-118.
Translator's note.)

and thereby solve, first the isomorphism problem, then the transformation problem.

Let G be a group with generators $S_1, S_2, \ldots, S_m$ and the
relations $R_1 = R_2 = \ldots = R_n = 1$, in which each generator appears
at most twice altogether in the relations.

1. If a generator S_1 appears in a relation, say $R_1 = 1$,
only once, i.e. if R_1 is equal to $S_1 T$, perhaps by cyclic
interchange of factors, where S_1 does not appear in T, then G
is isomorphic to a group G' with generators $S_2, \ldots, S_m$ and
relations $R_2 = \ldots = R_n = 1$, in which S_1 is replaced by T^{-1}.
$S_2, \ldots, S_m$ then appear at most twice in each relation. Thus by
this reduction one can reach the stage where each generator appears
in at most one relation, and in this one twice. We can therefore
begin by confining ourselves to groups having only one relation,
in which each generator appears twice.

2. Let G be a group with a single relation $R = 1$ in which
the generators $S_1, S_2, \ldots, S_m$ each appear twice. If we then
construct the group diagram, in which we attach to a point all the polygons
corresponding to cyclic interchanges of R, then in general we do not
find the same simple situation as we did for the fundamental groups
of one- and two-sided surfaces. On the contrary, the polygons
will come together in several sheets.

E.g. if we take the group

$$G \begin{cases} \text{generators} & S_1, S_2, S_3, S_4 \\ \text{relation}: & S_2 S_1 S_1 S_3 S_4 S_4 S_2 S_3 = 1 \end{cases}$$

then the eight associated polygons fall into two classes. The four polygons which join the same-sensed pairs of edges $S_2 S_1$, $S_1^{-1} S_1^{-1}$, $S_1 S_3$, $S_3^{-1} S_2^{-1}$ on the point constitute one sheet, while those for the edge pairs $S_3 S_4$, $S_4^{-1} S_4^{-1}$, $S_4 S_2$, $S_2^{-1} S_3^{-1}$ constitute the other. If we now transform this group by introducing a new generator $\Sigma_1 = S_1 S_3$ we obtain

$$G' \begin{cases} \text{generators}: & \Sigma_1, S_2, S_3, S_4 \\ \text{relation}: & S_2 \Sigma_1 S_3^{-1} \Sigma_1 S_4 S_4 S_2 S_3 = 1 \end{cases}$$

The polygons of the group diagram attached to a point again fall into two classes : those with the edge pairs $S_2 \Sigma_1$, $\Sigma_1^{-1} S_3$, $S_3^{-1} S_2^{-1}$ resp. $\Sigma_1 S_4$, $S_4^{-1} S_4^{-1}$, $S_4 S_2$, $S_2^{-1} S_3^{-1}$, $S_3 \Sigma_1^{-1}$. If one introduces 2 more new generators $S_2 \Sigma_1 = \Sigma_2$ and $\Sigma_1^{-1} S_3 = \Sigma_3$ one obtains the group

$$\begin{cases} \text{generators}: & \Sigma_1, \Sigma_2, \Sigma_3, S_4 ; \\ \text{relation}: & \Sigma_2 \Sigma_3^{-1} S_4 S_4 \Sigma_2 \Sigma_3 = 1. \end{cases}$$

One sees that the generator Σ_1 does not appear in the relation at all, and the six polygons at a point in the group diagram corresponding to the relation constitute a single sheet. In fact the series of edge pairs is $\Sigma_2 \Sigma_3^{-1}$, $\Sigma_3 \Sigma_2$, $\Sigma_2^{-1} \Sigma_4^{-1}$, $\Sigma_4 \Sigma_4$, $\Sigma_4^{-1} \Sigma_3$, $\Sigma_3^{-1} \Sigma_2^{-1}$.

Thus if we omit the generator Σ_1 we obtain a group whose diagram is a net of regular 6-gons in the hyperbolic plane.

The method explored in this special case may be made general, as we shall show in the following.

Suppose that

$$S_1 S_2, \; S_2^{-1} S_3, \ldots, \; S_\ell^{-1} S_1^{-1}$$

are ℓ edge pairs belonging to polygons which constitute a single sheet $\mathcal{B}$. Each of the generators $S_1, \ldots, S_\ell$ appears among the pairs either twice or four times. We shall first consider the two extreme cases :

a) The generators $S_1, \ldots, S_\ell$ all appear four times in the edge pairs of the sheet $\mathcal{B}$: since each generator appears in the relation at most twice, by hypothesis, then it belongs to at most four edge pairs. Thus none of the generators $S_1, \ldots, S_\ell$ forms an edge pair with a generator different from any of these. But since there is only one relation by hypothesis, and hence only one simple polygon in the group diagram, in which each generator appearing in the relation is represented by edges , then no generators other than $S_1, \ldots, S_\ell$ appear in the relation. These generators alone, in the presence of the relation, determine a group whose diagram is a net of regular n-gons, where $n = 2\ell$.

b) All generators $S_1, \ldots, S_\ell$ appear twice in the edge pairs of a sheet $\mathcal{B}$: we choose as new generators

$$\Sigma_2 = S_1 S_2, \ \Sigma_3 = S_2^{-1} S_3, \ldots, \Sigma_\ell = S_{\ell-1}^{-1} S_\ell$$

whence

$$S_\ell^{-1} S_1^{-1} = (\Sigma_2 \Sigma_3 \ldots \Sigma_\ell)^{-1}.$$

All generators $S_1, \ldots, S_\ell$ may be expressed in terms of $S_1, \Sigma_2, \ldots, \Sigma_\ell$:

$$S_1 = S_1, \ S_2 = S_1^{-1} \Sigma_2, \ S_3 = S_1^{-1} \Sigma_2 \Sigma_3, \ldots \ .$$

Thus the given group G is isomorphic to a group Γ in which $S_1, S_2, \ldots, S_\ell$ are replaced by $S_1, \Sigma_2, \ldots, \Sigma_\ell$ according to the above formulas. By hypothesis each generator S_i $(i = 1, \ldots, \ell)$ appears in the relation only twice, once in the context $S_{i-1} S_i$, the second time in the context $S_i^{-1} S_{i+1}$, but by the above these contexts are identical with $\Sigma_i \Sigma_{i+1}$, and for $i = \ell$, with $(\Sigma_2 \Sigma_3 \ldots \Sigma_\ell)^{-1}$. Thus by introduction of the new generators, S_1 no longer appears in the relation. Thus by this transformation we have reduced the number of generators appearing in the relation by one (also, as is easy to see, the number of sheets formed by the polygons at a point of the group diagram has also become one smaller). In addition, after the transformation each generator appears in the relation either twice or not at all.

General case : Some generators appear four times and some twice in the sheet-forming pairs. Then there is a pair consisting of generators which appear four times and twice respectively, since any two successive pairs have a generator in common. We may assume that this is the case for the first pair $S_1 S_2$ and in fact that S_1 appears four times, S_2 twice. If we step again through the series of pairs, then before the end we shall reach a pair in which S_1 appears again, and in fact the first such pair has the form $S_r^{-1} S_1$; for if it had the form $S_r^{-1} S_1^{-1}$ the first r pairs would already constitute a sheet and S_1 would occur only twice among the pairs. The series therefore runs

$$S_1 S_2, \ S_2^{-1} S_3, \ldots, S_r^{-1} S_1, \ S_1^{-1} S_{r+1}, \ldots, \ S_\ell^{-1} S_1^{-1}$$

and contains $\ell+1$ pairs. We now set

$$S_1 S_2 = \Sigma_1$$

and introduce Σ_1 as a new generator in place of S_1. We now obtain from the pairs the series

$$\Sigma_1, \ S_2^{-1} S_3, \ldots, S_r^{-1} \Sigma_1 S_2^{-1}, \ S_2 \Sigma_1^{-1} S_{r+1}, \ldots, S_\ell^{-1} S_2 \Sigma_1^{-1}$$

and this yields the ℓ sheet-forming pairs

$$S_2^{-1} S_3, S_3^{-1} S_4, \ldots, S_r^{-1} \Sigma_1, \Sigma_1^{-1} S_{r+1}, S_{r+1}^{-1} S_{r+2}, \ldots, S_\ell^{-1} S_2.$$

Here Σ_1 appears only twice, and we have thus converted our sheet into one with a pair less. The relation between the new generators, when S_1S_2 is replaced by Σ_1, again contains each generator either twice or not at all. For Σ_1 appears twice : in place of S_1S_2 as well as the substitute edge $\Sigma_1 S_2^{-1}$ for S_1 where it appears in the relation for the second time. Similarly S_2 appears twice, namely in this substitute edge , as well as the place where it appeared for the second time originally. It is worth remarking that the above conclusion is no longer correct when S_2 also appears four times among the original pairs; for if the original relation ran, say,

$$S_1 S_2 S_t \ldots = 1$$

then after insertion of Σ_1 the pair $S_t^{-1} \Sigma_1^{-1}$ would follow $S_t^{-1} S_2^{-1}$, after which $\Sigma_1 S_2^{-1}$ and $S_2 S_u$ etc. would follow, i.e. we would again have $\ell+1$ sheet-forming pairs, as before the transformation. One can easily be convinced of this phenomenon by an example.

Using this process one can successively diminish the number of pairs of a sheet until one comes to the extreme case b), i.e. until there are the same number of pairs as generators corresponding to the sheet. But then we can use the separate reduction procedures above, by means of which one sheet disappears, and one generator no longer appears in the relation. Then we admit more relations again and repeat until we arrive at the following normal form of the group : among the generators are w generators which appear in no relation ("free generators") and v classes of other generators; the generators of each class appear in only one relation, and in this relation, twice. The generators of each class, together with the class relation, determine a group whose diagram is a regular net in the euclidean or non-euclidean plane (see p.156).

If the given group is already determined by a single relation then indeed arbitrarily many further free generators can be given, however, only one sheet-forming class. In this case, as is easy to see, the group is the fundamental group of curves on a surface complex consisting of a closed surface and w non-bounding closed curves attached to a point of the latter. On the other hand, the original form of the group relation leads to surfaces with self-contact points. In place of free curves we can also introduce "horns" (Fig. 5), i.e. spheres with a self-contact point. The different forms of surface corresponding to the group then correspond to the fact:

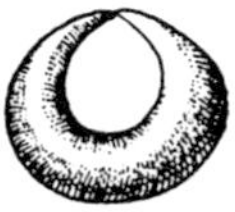

Fig. 5.

Each closed surface with w self-contact points is homeomorphic to a singularity-free surface with w horns attached.

In the general case, i.e. when the group is determined by arbitrarily many relations, the group in the reduced form is the fundamental group of v closed surfaces and w free closed curves attached to the same point. The group in its original form corresponds to the system in which some of the free curves are transformed into self-contact points of the surface.

For each group of the kind considered, i.e. in which each generator appears at most twice in the defining relations, we thus have a characteristic series of numbers:

$$p_1, \ldots, p_{\nu_1}, \; k_1, k_2, \ldots, k_{\nu_2}, \; w.$$

Here $p_1, \ldots, p_{\nu_1}$ denote the genus numbers of the two-sided surfaces corresponding to the generator classes of the given group, $k_1, \ldots, k_{\nu_2}$ the characteristics of the associated one-sided surfaces, and finally w is the number of free curves. Since homeomorphic surface complexes have isomorphic fundamental groups we have the theorem:

   If the characteristic numbers of two groups coincide, then these groups are isomorphic.

The converse also holds.

   If two groups of the kind in question are isomorphic, then their characteristic numbers coincide (i.e. one can bring such a group to reduced form in only one way).

   In fact : let G and G' be the two isomorphic groups, each in reduced form, let M_2 be any closed manifold whose fundamental group corresponds to a generator class of G. We shall show that the elements of G' corresponding to this fundamental group are again all the elements of the fundamental group of a surface, or else

result from the elements of such a group by conjugation with a
certain element of G'.   For this purpose we remark that the group
diagram for the fundamental group of a complex of two or more surfaces
consists of infinitely many sheets, with two or more meeting at
every vertex.   Now let T be any element of the subgroup of G'
in question, then the different powers of T will only lie on the
same sheet if T either lies on a single sheet, or is transformable
into such an element.   Further : if T, U are two elements of the
subgroup, then the different powers of T, U and TU will only
belong to the same sheet if T and U are both conjugates of elements
on the same sheet.   Now since this subgroup of G' is isomorphic
to the fundamental group of a closed surface then by the theorem
proved on p.158 its representation in the group G' has a connectivity
of the same number of sheets.   This is not the case, as is easily
seen, if the different powers of T, U and TU belong to infinitely
many different sheets.   Thus the subgroup in question is itself a
subgroup of a fundamental group of a surface corresponding to G'
and since we can apply our conclusion just as well to the image of
a fundamental group corresponding to G' in G, it follows that
each fundamental group of a surface associated with G corresponds
to a fundamental group of a surface associated with G', possibly
transformed by an element of G'.   Then it follows from the earlier
theorem that only the fundamental groups of homeomorphic surfaces
can be isomorphic,   that the numbers $p_1, \ldots, p_{\nu_1}, k_1, k_2, \ldots, k_{\nu_2}$

are the same for G and G'. Finally, it follows from the Tietze theorem that the horn number w must also be the same for G and G'.

After we have brought the groups in question into a canonical form it is easy to give a solution to the transformation problem. Let S_1 and S_2 be two elements of one of these groups, K_1 and K_2 the corresponding curves on the associated surface complex, which in the canonical form we have developed consists of one- and two-sided surfaces, plus horns, which all meet at a <u>single</u> point. S_1 and S_2 are then transformable into each other iff K_1 and K_2 are continuously transformable into each other on the surface complex. K_1 may be divided into successive closed curves $L_1^{(1)}, L_1^{(2)}, \ldots, L_1^{(m_1)}$, where we assume that none of the curves $L_1^{(i)}$ is contractible to a point and also that no two successive curves $L_1^{(i)}$ and $L_1^{(i+1)}$ or $L_1^{(m_1)}$ and $L_1^{(1)}$ lie on the same surface of the complex. Likewise when we divide K_2 into $L_2^{(1)}, L_2^{(2)}, \ldots, L_2^{(m_2)}$. Then if K_1 and K_2 are to be transformable into each other, m_1 and m_2 must be equal. Further, if $m_1 = m_2 = 1$, $L_1^{(1)}$ and $L_2^{(1)}$ must lie on the same surface and be continuously transformable into each other there, and if $m_1 = m_2 > 1$, then with possibly a suitable cyclic rearrangement, the m_1 curves $L_1^1 (L_2^{(1)})^{-1}, L_1^2 (L_2^{(2)})^{-1}, \ldots, L_1^{m_1} (L_2^{(m_2)})^{-1}$ must be contractible to a point. The conditions are also sufficient, as well as necessary, as is easily seen.

Chapter III

Higher groups

§1. General Remarks

For each given infinite group one can find a four-dimensional (homogeneous) manifold whose fundamental group is isomorphic to the given group : each surface complex can be represented without singularities in a four-dimensional space.   If the surface complex has the given group as fundamental group then the four-dimensional neighbourhood (domain) of the complex likewise has the given group as fundamental group.   One can also find a closed four-dimensional manifold with these properties, and indeed in the four-dimensional neighbourhood of the singularity-free representation of the surface complex in five dimensions.   Two M_4 with the same fundamental group need not be homeomorphic.   For example: the four-dimensional neighbourhood of an elementary piece of surface and a spherical surface in five-dimensional space both have the identity as fundamental group, but they are certainly not homeomorphic.   For the second Betti number is equal to 1 for the first manifold, 3 for the second.

In the preceding chapter we have treated those groups in which each generator appears at most twice in the defining relations. It is easy to show that each group can be brought into a form where each generator appears at most three times in the defining relations.

In fact : let S be a generator which appears h times (h > 3) in the defining relations, then we introduce the relation

$$S = S_1$$

where S_1 is a new generator and replace S twice in the relations by S_1. Then S appears (h-1) times in the new relations, and S_1 three times. If h > 4 then we also take the relation

$$S = S_2,$$

replace S twice by S_2, and so on, until, with the aid of h-3 new generators $S_1, S_2, \ldots, S_{h-3}$ the given group is brought into a form where there is one fewer generator which occurs more than three times in the relations. In this way our assertion is proved.

The next simplest class of groups after that we have already treated is that comprising all groups in which the defining relations include only one generator which appears three times. We shall deal with this in a later work.

§2. Knot groups

As we have shown earlier[*)] the group of the trefoil knot is presentable in the following form:

$$\left\{ \begin{array}{l} \text{generators} : S_1, S_2, S_3, S_4 \\ \text{relations:}\ S_1 S_4^{-1} S_2 = S_2 S_4^{-1} S_3 = S_3 S_4^{-1} S_1 = 1. \end{array} \right.$$

[*)] Math. Ann. 69.

In the same place we presented the group diagram of this group and thereby solved the identity problem. Here we shall show that the group diagram can be represented by a regular net in a non-euclidean space. In fact the group diagram found consisted of parallel strips (see Fig. 11 of the work cited), three of which meet at each boundary. Now we construct a system of segments in the hyperbolic plane with the following properties: it contains no closed polygon, and emanating from each vertex there are three equal segments at equal angles $\frac{2\pi}{3}$. This is easy to obtain if one takes the lengths of the sides to be the base of an isosceles triangle with angle 0 at the top and base angles $\frac{\pi}{3}$. It then follows that a path of segments of this length and corner angles $\frac{2\pi}{3}$ does not close or cut itself (for the vertices lie on a horocycle), and hence also a path of segments of this length and corner angles $\frac{2\pi}{3}$ or $\frac{4\pi}{3}$ does not end or cut itself. Consider for example the hyperbolic metric generated by a fundamental circle in the euclidean plane. Over this circle we erect a right cylinder, and likewise vertically over each segment a parallel strip, and place points on the boundaries of the latter at equal distances so a point on one side of the boundary is perpendicularly opposite the mid-point of two successive points on the other side. We then connect each point on the boundary with the two nearest to it on the opposite boundary and we then have in this collection of segments on the parallel strips the same

infinite line segment complex which, when correctly labelled, is
the diagram for the group of the trefoil knot.   As a result we
have shown that this group is isomorphic to a (properly discontinuous)
group of   linear transformations of space, which transforms the
cylindrical surface into itself.

The transformation problem for the group in question is also
easy to solve.   For this purpose it is unnecessary to bring the group of
motions of the non-euclidean space with the cylinder as fundamental
figure into a canonical form : the   edge   path corresponding to
an element of the group traverses a series of parallel strips.
This is then only diminished by transformation if the projection of
the   edge   path on the non-euclidean plane contains doubled segments
running in opposite directions, including the case when the last term
is placed at the beginning.   One can always convert the edge
path by transformation into one which traverses the minimal number
of parallel strips.   Two edge   paths transformed into this form
are then transformable into each other iff they have the same initial
and final points,possibly after cyclic interchange of terms.   In this
way we have established a method for deciding the transformability of
two elements in a finite number of steps.

By quite analogous considerations one solves the transformation
problem for the knot groups corresponding to the curves closely

related to the trefoil knot, which are presented on p. 164 of
Math. Ann. 69[*]. The basis for the diagrams of these groups
again comes from infinite regular line segment complexes in the
hyperbolic plane without closed curves, of the type where five
resp. seven segments meet at each vertex. The solution of the
transformation problem follows exactly as for the trefoil knot.

It is also possible to give the solution of the transformation
problem for the fundamental groups of the Poincaré spaces associated
with these knots. Among these are infinitely many corresponding
to the trefoil knot. The transformation problem for these is
solvable without further investigation. The group diagrams of
the others admit representations on a complex of prisms, the cross-
section of which is a regular polygon net in the hyperbolic plane.
Thus the groups may be represented by real linear transformations
of   space in which a cylindrical surface is mapped into itself.
A motion of the group diagram into itself corresponds to a motion
of the non-euclidean plane determined by two corresponding points
and two corresponding polygons attached to these points. Any
two points A and A_1 of the group diagram generate two distinct
such motions   , namely those which carry A into A_1 and
A_1 into A. Now let S be an element for which we
want to decide whether it is transformable into an element T or not.
An arbitrary point O of the group diagram may be sent to A by S, to
B by T. With these two motions a point $\mathbf{0}$ of the cross-section

---

[*] p. 121 this volume (Translator's note).

net goes into $\mathcal{U}$ resp. $\mathcal{Y}$ and at the same time the net polygon $P_{\mathfrak{y}}$ attached to $\mathcal{O}$ goes into the net polygon $P_{\mathcal{U}}$ resp. $P_{\mathfrak{Y}}$. If S and T are transformable into each other these two net motions must be congruent and indeed transformable into each other by a net motion corresponding to a motion of the group diagram into itself. This may be decided in a finite number of steps by investigation using the methods developed in Chapter I. However, this condition is not yet sufficient. Suppose U is an element whose corresponding net motion carries the net motion corresponding to S into that of T. The edge path corresponding to the element $U^{-1}SU$, whose associated net motion is identical with the motion $\{ \mathcal{O} \mapsto \mathcal{Y}$, $P_{\mathfrak{y}} \mapsto P_{\mathfrak{Y}} \}$ may start from a point O_1 and end at a point B_1. (O, O_1 and $\mathcal{O}$ and similarly B, B_1 and $\mathcal{Y}$ each lie on a straight line perpendicular to the cross-section net.) Then S is only transformable into T when the segments OO_1 and BB_1 are equal and in the same direction.

In this way we have brought the necessary and sufficient conditions for transformability of two elements of the group considered into such a form that one can decide in a finite number of steps whether or not they are satisfied. At the same time we have made it possible to settle the question whether two given curves of the Poincaré spaces in question are continuously transformable into each other (with self-intersections).

§3. Groups with two generators

To solve the three fundamental problems for all groups with two generators (say a and b) still appears to be very difficult. This class includes all knot groups,* as is easily seen by geometric considerations. It is simple to settle the problem for those groups in which a appears in the defining relations three times and b twice. Difficulties already appear when a and b both appear three times in the defining relations. Here we shall only say a little about the group given by

$$a^3 b^3 = 1.$$

We construct the group diagram from "cubic hexagons", i.e. spatial hexagons whose edges are edges of a cube such that each face of the cube contributes two edges. We denote the edges of a cubic hexagon, with a given sense, by a, a, a, b, b, b. In order to construct the group diagram we associate the cubes corresponding to the cubic hexagons with prisms. Each prism is cut by a second prism in each of its cubes. Going through two successive cubes of a prism there are two further prisms, whose axes are perpendicular. This completely determines the group diagram. However, it is important to notice that identical points of the group diagram are only those which are so as a result of the relations between the prisms, not those whose coincidence results from the special metric properties of euclidean space. E.g. Two parallel prisms P_1 and P_2

---

* This is incorrect (Translator's note).

go through two cubes W_1 and W_2 of a prism which are separated by an odd number of cubes. If W_1' and W_2' are two cubes on P_1 resp. P_2 which lie on homologous sides of W_1 and W_2 separated by the same odd number of cubes, then in euclidean space the two prisms cutting P_1 resp. P_2 in W_1' and W_2' naturally coincide. However, in our group diagram they have no point in common. Thus our group is not isomorphic to a group of motions in euclidean space, but one can show on the other hand that it is isomorphic to a group of motions in hyperbolic space, admittedly not properly discontinuous.

With the exhibition of the group diagram the identity problem is immediately solved. The transformation problem may also be settled with the help of general principles which will be developed in a later work.

The next paper contains Dehn's last word on the conjugacy problem for surface groups, as well as a very simple solution of the word problem. The solutions, by what is known as Dehn's algorithm, are based on a detailed combinatorial study of closed paths in the group diagram. This study shows, in effect, that if a word for an element of the fundamental group can be reduced in length, then it can be reduced monotonically, namely by replacing any subword which is more than half the defining relator by its complement, and by trivial cancellations. (When only a conjugate of the original element is sought, the word is regarded as circular in applying these operations.)

A word equals 1 if and only if its reduced form is identical to 1, and two words are conjugate if and only if their reduced forms are identical as circular words (apart from some simple exceptional cases). Thus, just as Dehn's previous algorithm depended on the existence of the hyperbolic metric but used only the combinatorial structure of the group diagram, the new algorithm depends on the existence of the group diagram but uses only the structure of words, thereby achieving great computational efficiency.

The geometrical interpretation of the algorithm for the

word problem is particularly simple : a closed edge path
in the group diagram can be contracted to a point by trivial
cancellation and the operation of pulling a subpath which
traverses more than half a polygon boundary to the other side
of the polygon in question.   To prove this one only has to
show that a closed path which admits no cancellation contains
a subpath traversing more than half a polygon boundary, and
Dehn's argument for this is extremely simple.

His argument for the conjugacy problem is more complicated,
but along the same elementary lines.   More sophisticated
proofs, using the Euler characteristic, are in Reidemeister
[1932] and Lyndon & Schupp [1977].

Dehn's algorithm for the conjugacy problem and its proof,
being completely combinatorial, avoid the axis and distance
curves which were central to the earlier proofs.   But
surprisingly, the ghost of the axis is still there.   Nielsen
[1927, §7g] points out that applying Dehn's algorithm to a
word c amounts to transforming a c-chain (cf. introduction
to the previous paper) into a chain which lies as close as
possible to the axis of the motion which maps the c-chain
onto itself.   This observation clarifies the exceptional

cases in Dehn's proof — they correspond to different chains which lie equally close to the axis — and suggests refinements of the algorithm reducing the number of exceptions which have to be considered (Nielsen [1927, §7g]).

It is also suggested, though Nielsen does not say so, that we re-examine the original application of the axis (Poincaré [1904]), which was to the simple curve problem; deciding whether a given curve is homotopic to a simple curve. Recall, from the introduction to Dehn's lecture notes on surface topology (especially Fig. 2 there), that deciding this question reduces to finding the course of the axis through the cells of the group diagram. The approximation of the axis by an edge path determines the course of the axis up to a finite (though possibly large) number of topologically distinct possibilities. These possibilities can then be examined in turn to see whether any of them yields a simple curve. This may explain the claim in Dehn [1922] that Poincaré's algorithm for the simple curve problem can be greatly simplified. Recently, Birman and Series have used Nielsen's methods to develop this idea into an efficient algorithm for the simple curve problem.

REFERENCES

M. Dehn [1922] : Über Kurvensysteme auf zweiseitigen Flächen, mit Anwendung auf das Abbildungsproblem (Breslau lecture notes).

R.C. Lyndon and P.E. Schupp [1977] : Combinatorial Group Theory, Springer-Verlag.

J. Nielsen [1927] : Untersuchungen zur Topologie der geschlossenen zweiseitigen Flächen, Acta Math., 50, 189-358.

H. Poincaré [1904] : Cinquième complément à l'analysis situs, Rend. circ. mat. Palermo 18, 45-110.

K. Reidemeister [1932] : Einführung in die kombinatorische Topologie. Teubner, Leipzig.

TRANSFORMATION OF CURVES ON TWO-SIDED SURFACES

The problem we shall deal with in what follows is one of the simplest of topology : given two closed curves on a closed two-sided surface, to decide whether one may be "transformed" into the other by a continuous deformation. The solution of this problem for surfaces of genus $p > 1$ by means of "polygon groups" and hence on the basis of the metric of the hyperbolic plane is evident, and is indicated e.g. by Poincaré (Rend. Circ. Mat. Pal. 1904), also developed more precisely by me in Math. Ann. 71*. In the same work I have given a method for deciding the question purely topologically without the help of the metric. However, in the foundation of this method I have made essential use of properties of figures in the hyperbolic plane. For surfaces of genus $p = 0$ and $p = 1$ the solution of the problem is very simple : in the first case all curves are transformable into each other, in the second case the fundamental group is abelian, and each curve is transformable into one which traverses a fixed curve C m times and a fixed curve Γ μ times, and these numbers m and μ are independent of the particular transformation, so that the

---

*Cf. the analytic exposition in the dissertation of Gieseking, soon to appear.

transformation problem is solved.

It is now a curious fact, and characteristic of the imperfect nature of topological investigation, that the solution of the transformation problem for $p > 1$ was previously unknown, despite its simplicity in the case $p = 1$: if one dissects the surface by $2p$ cuts $a_1, b_1, \ldots, a_p, b_p$ into a simply connected surface piece, the boundary of which is formed by the series of curves $a_1, b_1, a_1^{-1}, b_1^{-1}, a_2, \ldots, b_p^{-1}$, then each closed curve L on the surface may be transformed into one "generated" by these curves, say

$$a_1^{\alpha_1} \ldots b_p^{\beta_p} a_1^{\alpha_1'} \ldots \equiv A.$$

If there is, then, possibly after a cyclic interchange of terms, a part of the expression A which contains more than $2p$, say q, terms of the relation

$$a_1 b_1 a_1^{-1} b_1^{-1} \ldots b_p^{-1} = 1$$

in order or reverse order, then one replaces this part by the equal expression consisting of the $4p-q$ remaining terms. Furthermore, one cancels any successive terms C and C^{-1} which result, possibly after cyclic interchange, and continues these two processes as long as possible. Then A is finally converted into an expression K which we shall call a reduced expression. K represents a curve into which L is transformable. Then we have

Main Theorem : <u>Apart from some (easily settled) exceptional cases, each curve determines a unique reduced expression, up to cyclic interchange.</u>

<u>Apart from the exceptional cases, two curves are transformable into each other if and only if their reduced expressions are identical up to cyclic interchange.</u>

A few exceptional cases are very obvious, e.g. two expressions consisting of half the terms of the relation are transformable into each other, for $p = 2$

$$a_1 b_1 a_1^{-1} b_1^{-1} \quad \text{and} \quad b_2 a_2 b_2^{-1} a_2^{-1}.$$

Less trivial exceptional cases will be dealt with later when the occasion arises.

In what follows we assume a knowledge of the simplest properties of the mapping of the two-sided surface of genus $p > 1$ onto a net of $4p$-gons, $4p$ of which meet at each vertex. This may be found in Chapter I, §1, paragraphs 1-5 and §5 of

my work in Math. Ann. 71.   I recall here only that this net
is the diagram of an infinite group, the fundamental group of
the surface, generated by $a_1, b_1, \ldots, b_p$ with relation

$$a_1 b_1 a_1^{-1} b_1^{-1} \ldots b_p^{-1} = 1.$$

All curves on the surface which are transformable into each
other correspond to conjugate elements of the group.   In particular,
curves which are contractible to a point, the identity element
of the group, are represented by closed polygonal paths of
the net.

§1.   The identity problem

Theorem 1 : <u>If the element</u> $a_1^{\alpha_1} b_1^{\beta_1} \ldots a_1^{\alpha_1'} \ldots$ <u>of</u>
<u>the fundamental group equals the identity, then this expression</u>
<u>is reducible.</u>   I.e. there is either a sequence of q terms
(q > 2p) in the expression which also occurs in the defining
relation, and hence may be replaced by the remaining 4p-q
terms, or else two successive terms s and s^{-1} occur in the
expression and therefore may be cancelled.

To prove this we have merely to show that each closed
polygonal path in the group diagram of 4p-gons has more than

2p edges in common with a net polygon, or else has two
oppositely directed edges occurring in succession.

  We consider first the complex G_1 consisting of a
single net polygon, then the complex G_2 of net polygons
which have a point in common with the boundary of G_1, then
G_3, which results from G_2 by adding all polygons which have
a point in common with the boundary of G_2, etc. .   Each closed
polygonal path in G_1 certainly has the required property,
but so also does each closed path in G_2.   In fact, each vertex
of G_2 which does not belong to G_1 is a boundary point of
the simply connected complex G_2, and at most one edge from it
goes to the interior, in which case there are two reducible
polygonal paths emanating from the point which belong wholly
to the boundary.   Thus a closed polygonal path which possesses
no successive oppositely directed edges is either wholly
contained in G_1 or necessarily contains a reducible polygonal
path on the boundary of G_2.   Exactly the same argument allows
us to carry the proof from G_2 to G_3, G_4, etc., whence our
theorem is proved.

  Our proof makes essential use of the properties of the
group diagram, whereas the theorem itself yields a process

for deciding directly whether an expression represents the
identity element, without construction of the group diagram.
This is an advantage over the method in which one first constructs the group diagram and then uses it to decide whether
an expression yields a closed polygonal path or not.

For what follows it is necessary to find further properties
of closed polygonal paths : we shall _first_ confine ourselves
to _singularity-free paths_.   A series of net polygons
$P_1, P_2, \ldots$, where each has an edge in common with its successor,
but no two polygons have a common element apart from this, will
be called a _chain_.   We begin our considerations, for simplicity,
with a polygon contained in the region Γ bounded by the path
Π in question and having only a _single_ open polygonal path in
common with Π.   Using the above notation, we denote this
polygon by G_1.   If Π does not merely consist of the boundary
of a polygon, then polygons of G_2 must be contained in Γ, for
otherwise Π would have double edges.   However, if only _one_
polygon of G_2 is in Γ then this polygon and the first
constitute a chain on the boundary of which there are two
reducible paths of 4p-1 segments; while if several polygons
of G_2 are in Γ, the boundary of the region composed of G_1

and these polygons has at least three reducible paths of at least $4p-2$ segments. Now since a polygon of G_3 never has two edges in common with the boundary of G_2, otherwise two polygons of G_3 would simultaneously have a segment in common with each other and with G_2, none of the reducible paths of Γ can be destroyed without an equal number taking their place. By continuing this argument we arrive at the result:

A closed singularity-free curve of the net has at least three reducible paths of at least $4p-2$ edges with the single exception of the case where it constitutes the boundary of a chain or of a single polygon.

We now consider curves with singularities, firstly those with only one. For each such curve Π we can begin at the singular point and traverse the curve Π in such a way as to describe a simple closed curve (a loop). This loop then has no more than two reducible polygonal paths if and only if it is the boundary of a chain; because of the fact that Π may not end at the singular point, as the loop does, this means at most one reducible polygonal path of the loop does not lie on Π itself. But now each curve with singularities has at least two loops without a common edge. Thus in order for Π to

have less than three reducible paths of at least 4p-2 edges,

Π must have two loops without a common edge, and these must

each bound a chain.   But it then follows also that neither of

the two loops can have singular points.   For if I consider

a loop ABA, where the point B is a singular point of Π,

then two cases are possible :

   (1) (Fig. 1) The two paths from

A other than the loop run to points

B and C on the loop without them-

selves forming loops, i.e. without

double points; let these paths be,

say, AXB and AYC.   If AXB

Fig. 1.

together with BA bounds a chain, then this has BA in common

with the loop ABCA, so that BA consists of a single edge.

Thus among the reducible paths on AXBA at least one with at

least 4p-1 segments remains for the path AXB on Π.   But

if AXB together with BA bounds no chain, then indeed BA

need not be a single edge, but in return AXBA has at least

three reducible paths of 4p-2 edges, of which at least one

lies on the path AXB or on the path AB of Π.   By considering

the other path AYC, it follows that there is at least one

reducible path of at least 4p-2 edges on it, or on AC. Then when we bear in mind that the loop AYCBA as well as the loop AXCBA has at least one reducible path of Π with 4p-2 edges, it follows immediately that in this case there are again three reducible paths of at least 4p-2 edges on Π.

2. (Fig. 2) One of the paths on Π from A other than the loop ABA describes a loop CYC before returning to the singular point on ABA. Then if Π is not to have three reducible paths, the other path from A must run singularity-free to the singular point B of the loop ABA (otherwise one would have at least three separated loops); however, its presence results in at least two reducible paths of 4p-2 edges on Π, as above, so that together with the path yielded by the loop CYC we again have three such paths for Π.

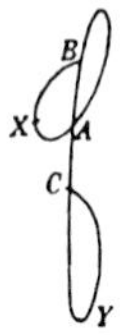

Fig. 2.

Thus the existence of a singular point B on the loop necessarily has the consequence that Π has at least three reducible polygonal paths of at least 4p-2 edges. If this

is not the case, then Ⅱ divides the plane into compartments
$Ⅱ_1$, $Ⅱ_2$, ..., $Ⅱ_m$ each of which has only a point in common with
its successor, and is either a chain or a single net polygon.
If we also admit double edges, then nothing essential is
changed in our argument. Summing up, we obtain:

Theorem 2. <u>A closed curve Ⅱ of the net has at least</u>
<u>three reducible polygonal paths of at least 4p-2 edges,</u>
<u>provided it is not the boundary curve of a chain, or a series</u>
<u>of chains (or net polygons) each of which has either a single</u>
<u>point in common with its successor, or else is connected to it</u>
<u>by a double edge of Ⅱ.</u>

Thus if Ⅱ does not have three reducible polygonal paths
of at least 4p-2 edges, and if it is not the boundary of a
<u>single</u> net polygon, then Ⅱ has at least two reducible polygonal
paths of at least 4p-1 edges.

§2. The transformation problem

With the help of Theorem 2 it is now easy to solve the
transformation problem, i.e. the problem of reduction of
closed curves on a surface of genus p > 1, in the way intended.

Let U and V be two reduced expressions in the notation of the introduction, i.e. after cyclic interchange U and V contain no reducible polygonal paths or paths of the form SS^{-1}. (This reduction may be made without use of the group diagram, by directly applying the fundamental relation to the given expression.) Now let U and V be transformable into each other, i.e. by suitable choice of a third expression T the polygonal path

$$TUT^{-1}V^{-1} = \Pi$$

is closed. By first making cyclic interchanges in the terms of U and V we can arrange that Π contains no path of the form SS^{-1} ; furthermore, by possible reduction of T and T^{-1} we can ensure that no reducible polygonal paths lie on these paths.

We assume now that Π <u>is not the boundary of a chain or a series of chains</u> (see §1). Then by §1, Π has at least three reducible polygonal paths of not less than 4p-2 edges. Since the paths U, V and T themselves contain no such paths, there must be two of these paths which have a part in common with U but none with V, or else two which have a part in

common with V but none with U.   We take, say, the first
possibility, and let tu and t'u' be the two reducible
polygonal paths, where t and t' belong to T and T^{-1}
respectively, and u and u' belong to U.   Since t and
t' are two final sections of the same path T, one of them
must be only <u>one</u> edge long.   For tu and t'u' must belong
to the boundary of a net polygon, however two edges appear in
the same order only once in the boundary of a polygon.   Thus
if t and t' both have more than one edge, u and u' must
have the same beginning.   But then the first and last edge
of U are the same but oppositely directed, and U can be
reduced by cancellation of a term ss^{-1} after cyclic interchange, contrary to hypothesis.   However, if t has only
one edge, u must have at least 4p-3 edges because tu has
at least 4p-2.   Thus U has a path of 4p-3 edges, which
is reducible because p > 1, and we have a contradiction to
the hypothesis that U is a reduced expression.

Consequently, we have the result:

<u>Two reduced paths which are transformable into each other</u>
<u>always constitute the boundary of a chain or series of chains</u>
<u>when a suitable transforming expression is added.</u>

For a general element, however, it is impossible to find elements T and V such that $TUT^{-1}V^{-1} \equiv \Pi$ is the boundary of a proper chain, on the contrary Π will consist of a double path traversing the same route in opposite directions.

In fact, if the boundary is not to be degenerate, each of U and V must, apart from double paths in Π, decompose into polygonal paths of not less than $2p-2$ edges which lie on the boundary of the same net polygon : in this case we have two reducible paths of at least $4p-1$ edges on Π. These may both lie simultaneously on U and V, in which case it follows as above that T consists of a single edge at most. Or else one lies, say, on TU and the other on $T^{-1}V^{-1}$. Let one be denoted by tu, the other by $t'v'$, in the sense used above. Then one can replace t in T by cu^{-1} (where c is a generator) or U^{-1} respectively, according as tu consists of $4p-1$ or $4p$ edges. But then we can leave out the term u^{-1} in the transforming expression by carrying out the corresponding cyclic interchange of terms in U, and obtain a new transforming expression $\overline{T}$ which has at least $2p-2$ terms fewer than T. This can be continued as long as T does not form a reducible polygonal path in combination with U and V. Thus we obtain the result :

Corollary to the Main Theorem : If U and V, together with a suitable transforming expression, form the boundary of a chain etc., then by suitable cyclic interchange of the terms in U and V one can find that U equals V or else U goes into V by transformation with a single generator.

This result suffices to settle the transformation problem in the exceptional case by reducing it to the identity problem of the previous paragraph. In fact, if U and V are transformable into each other we must have

$$c\bar{U}c^{-1}\bar{V}^{-1} = 1$$

or

$$\bar{U}\bar{V}^{-1} = 1$$

where c is any one of the 2p generators and $\bar{U}$, $\bar{V}$ are expressions resulting from U, V respectively by cyclic interchange. Thus we have now completely settled the transformation problem. The case we have called the exceptional case is in fact, as remarked above, quite special : in order for it to occur it is necessary that paths of length at least 2p-2 edges from a net polygon appear in U as well as V. Accordingly, we can immediately conclude e.g. : two expressions of the form $c_1^{m_1} c_2^{m_2} \ldots c_\nu^{m_\nu}$ where $c_1, c_2, \ldots$ are any generators and $|m_1|, |m_2|, \ldots, |m_\nu|$ are all greater than 1, are transformable into each other only when they result from each other by cyclic interchange of terms, likewise two expressions of the form

$$a_{n_1}^{\alpha_1} a_{n_2}^{\alpha_2} \ldots a_{n_q}^{\alpha_q}$$

or

$$b_{\ell_1}^{\beta_1} b_{\ell_2}^{\beta_2} \ldots b_{\ell_q}^{\beta_q}$$

in which only one or other half of the generators appear are transformable into each other if and only if they result from each other by cyclic interchange of terms. These are results which could have been derived only with great difficulty with the previous methods.

We give yet another example for the exceptional case :
let $p = 2$, then

$$a_1 b_1 a_1^{-1} a_1^{-1} b_1^{-1} a_2 = U$$

and

$$a_2 b_2^{-1} a_2^{-1} a_1^{-1} b_2 a_2 = V$$

are transformable into each other, for

$$b_2^{-1} U b_2 V^{-1} = 1$$

because the expression on the left-hand side is, in detail,

$$b_2^{-1}a_1b_1a_1^{-1}a_1^{-1}b_1^{-1}a_2b_2a_2^{-1}b_2^{-1}a_1a_2b_2a_2^{-1}.$$

Here the last three and the first four terms constitute a path of $4p-1 = 7$ edges of the fundamental polygon, and are thus replaceable by one edge, namely b_1. There then remain eight terms which exactly describe the boundary of a fundamental polygon.

The method for deciding transformability, as well as its proof, is quite free of metric elements. Indeed, as far as the special case of the identity problem goes, as well as Theorem 2, the particular labelling of the net edges is not used, so that these results hold for all groups whose diagrams are nets of $4p$-gons, $4p$ of which meet at each vertex. It is easy to see that even this much need not be assumed, but merely that the polygons are at least seven-sided, and that at least four polygons meet at each vertex.

In the solution of the transformation problem the special labelling of the net edges is used at <u>one</u> place, namely the property that a [labelled] polygon path of more than one edge

occurs at most once in a given net polygon.   However, this
property of the labelling follows from the topological property
of the net, that there is no vertex where only two polygons
meet.

Of all Dehn's results, perhaps the one most understandable
to a non-mathematician is the following: a left-hand trefoil
knot cannot be deformed into a right-hand trefoil knot. The
statement concerns everyday objects, and only the proof is
difficult. Dehn's proof is based on the observation that a
deformation of one trefoil knot to the other would induce an
automorphism of the knot group which associates certain geometrically
significant elements, namely "latitude" and "longitude" curves
on the neighbourhood torus of the knot, which determine an
orientation of the ambient space.

Dehn sticks to the presentation and group diagram he
developed in Dehn [1910], which is far from easy to handle,
and only his virtuosity in hyperbolic geometry carries him through.
A much easier solution of the basic group theoretic problem,
which I have included as an appendix below, was obtained
algebraically by Schreier [1924], based on the presentation
$\langle A,B \; ; \; A^2 = B^3 \rangle$. This presentation is immediate from Dehn's
relation

$$a_4^{-2} a_1 a_4 a_1 = 1,$$

i.e.

$$a_4^{-3} (a_4 a_1)^2 = 1,$$

by setting $a_4 a_1 = A$, $a_4 = B$. Schreier's argument in fact covers all groups $\langle A,B ; A^a B^b = 1\rangle$ where a, b are integers, and hence all torus knot groups. (For a beautiful proof that all torus knot groups are of this form, see Seifert & Threlfall [1934].) Moreover, for these groups Schreier solves the problem posed by Dehn at the end of I below, to find a presentation of the automorphism group of a given finitely presented group. Dehn's pregnant remark, that he would attack this problem for surface groups "in a work to appear shortly", had a long gestation period. His results eventually appeared in the 1920's and 1930's, partly in his own work and partly as reworked by Nielsen, Baer and Goeritz. See the next two papers in this volume, and their introductions, for more information.

The brief investigation of the automorphisms of the figure eight knot group in III was completed by Magnus [1931], who showed that the two automorphisms given by Dehn do in fact generate the outer automorphism group. According to Magnus [1979], the Kiel student who constructed the diagram of the figure eight knot group was named Fritz Klein. It seems likely that he died in World War I, and the diagram of the group has never appeared anywhere. It may have been based on the 1912 Münster dissertation of Dehn's student Hugo Gieseking (the relevant part of this thesis is expounded in Magnus [1974]). Gieseking arrives at the same group as a discontinuous group of motions of hyperbolic 3-space. This remarkable discovery became more understandable when Thurston [1977b] showed directly that the figure eight knot complement could be given a hyperbolic structure, and that this phenomenon is fairly typical of knot complements.

REFERENCES

M. Dehn [1910] : Über die Topologie des dreidimensionales Raumes. Math. Ann. 69, 137-168.

W. Magnus [1931] : Untersuchungen über einige unendliche diskontinuierliche Gruppen. Math. Ann. 105, 52-74.

    [1974] : <u>Noneuclidean Tesselations and Their Groups</u>. Academic Press.

    [1979] : Max Dehn. Math. Intelligencer 1, 132-143.

O. Schreier [1924] : Über die Gruppen $A^a B^b = 1$. Abh. Math. Sem. Univ. Hamburg 3, 167-169.

H. Seifert & W. Threlfall [1934] : <u>Lehrbuch der Topologie</u>. Teubner, Leipzig.

W. Thurston [1977b] : Geometry and topology of 3-manifolds. Lecture notes, Princeton Univ., Mathematics Department.

THE TWO TREFOIL KNOTS

In the following I wish to show that investigation of the groups associated with interwoven space curves (knots) puts us in a position to treat basic topological problems simply and rigorously. For this purpose I have chosen a seemingly special example, which on the one hand is of general interest, while on the other hand being so simple that no special difficulties stand in the way of the method.

The <u>trefoil knot</u> (Fig. 1) is the simplest knot, i.e. the simplest interwoven space curve which cannot be continuously transformed in space into a circle. It is the only knot with a plane projection having only three double points. In space one can distinguish <u>two different kinds</u> of trefoil knot, which result from each other by reflection, say, in a plane. Both kinds are shown in Fig. 1.

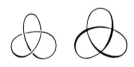

Fig. 1.

<u>Our problem is to find a way of distinguishing the two kinds of trefoil knot in space topologically</u>[*], i.e. to show that the left knot is not convertible into the right by a

---

[*] And in fact also in the closed spherical space, contrary to a remark in the Enzyklopädie.

continuous space transformation. Of course this fact is

already known to those who have gone into the subject of knots

and visualized space curves, say, as threads. But this knowledge,

while perhaps adequate practically, is very far from being a

real proof[**]. In any case, the phenomenon is very remarkable[***]

The space with a left trefoil knot and the space with a right

trefoil knot are homeomorphic, i.e. so composed that the knots

occupy corresponding regions in the two spaces. Despite this

the two knots in the same space are not convertible into each

other by a continuous transformation of the space. Thus we

have before us a kind of <u>topological symmetry</u>. In two-dimensional

space (on the disc) symmetric regions not convertible into each

other are far less beautiful, as we may note incidentally.

Fig. 2 shows two of them, chosen more

or less arbitrarily.

In what follows I shall now

lay the foundations of the proof.

The route will be clearer if one

first considers the continuous

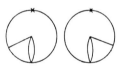

Fig. 2

transformations Δ of a three-dimensional manifold M_3 into itself under which two curves K_1 and K_2 lying in M_3 go into each other (in our case M_3 is either a simply connected piece of ordinary space or the whole three-dimensional spherical space).

If K_1 and K_2 go into each other under Δ then the groups* of K_1 and K_2 can be mapped isomorphically onto each other in such a way that all curves which correspond under Δ also correspond under the isomorphism.

This "main condition" follows immediately from the circumstance that any curve which bounds a disc in the complement space of K goes into a curve which bounds a disc in the complement space of K_2 under the continuous transformation Δ of M_3. Thus if $a_1 \ldots a_n$ are elements of the group of K_1 for which

$$a_1 \ldots a_n = 1$$

then for the corresponding elements of the group of K_2 we have

$$a_1' \ldots a_n' = 1$$

where a_1 and a_1' correspond to curves which go into each

---

*
See the remark below.

other under Δ .   Thus the correspondence $a_i \mapsto a_i'$ is an
isomorphic mapping of the group onto itself.   If we surround
a point of a curve K by a sufficiently small 3-cell E_3 in
M_3 then K will cut the spherical surface bounding E_3 in
two points A and B ; a curve β which separates A and B
on the boundary of E_3 will be called a <u>curve of latitude</u>.
If we surround K by a sufficiently narrow torus inside M_3
then this torus will have no point in common with K.   The
torus will contain a certain curve of latitude which does not
separate the surface.   A curve on the surface which cuts β
just <u>once</u> will be called a <u>curve of longitude</u> λ.   Then as a
special case of the main condition we have:

   a)   <u>The isomorphism between the groups of K_1 and K_2</u>
<u>can be chosen in such a way that each curve of latitude</u> β_1
<u>of K_1 goes into a curve of latitude</u> β_2 <u>of K_2, and likewise</u>
<u>a curve of longitude</u> λ_1 <u>of K_1 goes into a curve of longitude</u>
λ_2 <u>of K_2.</u> [*]

   In fact, Δ takes a point of K, resp. spherical neighbourhood
of a point of K_1 resp. torus around K_1 into a point of K_2
resp. spherical neighbourhood of a point of K_2 resp. torus

---

[*] See the remark at the end of this work.

around K_2, and K_2, like K_1 has two points in common with
the spherical neighbourhood, none in common with the torus, etc.

This condition a) holds for all two-sided and one-sided
manifolds M_3. <u>If M_3 is two-sided</u> we can sharpen condition
a) still further : if we take any pair β and λ for a curve
K and give β and λ a sense, then β gives, first to the
3-cell E_3 around a point P of K, and then to the whole manifold M_3, an indicatrix.   For example, one can associate with

E_3 the screw determined by the sense of

β and the sense of K when traversed in

the same direction as λ   (see Fig. 3).

Now since the indicatrix of a <u>two-sided</u>

M_3 is unaltered under continuous transformation

into itself, we have the following sharpening

Fig. 3

of the condition a).

a')   <u>The isomorphism between the groups of K_1 and K_2 may</u>
<u>be chosen in such a way that when a pair of sensed curves</u> $β_1$
<u>and</u> $λ_1$ <u>of K_1 goes into a pair of sensed curves</u> $β_2$ <u>and</u> $λ_2$
<u>of K_2 the indicatrix of M_3 determined by</u> $β_1$ <u>and</u> $λ_1$ <u>is the</u>
<u>same as that determined by</u> $β_2$ <u>and</u> $λ_2$.

In our case M_3 is ordinary space.   The group $G_{K\ell}$ of the trefoil knot in this space is already known from earlier investigations[**] . In the next section we investigate the isomorphisms of $G_{K\ell}$ onto itself, the <u>automorphism group of</u> $G_{K\ell}$. It will be shown that these may be very simply described : apart from the obvious "inner" automorphisms there is essentially only one "outer" automorphism. In this way we cover all possibilities of associating curves relative to the left trefoil knot with curves relative to the right trefoil knot, in such a way as to induce an isomorphism between the groups of these curves. It turns out that <u>the condition a') is never satis-fied</u>, so that our goal is achieved.

I have sought a further clarification of the problem by briefly treating an "amphicheiral" knot in the third section, i.e. an intertwined space curve not transformable into a circle, but convertible into its own mirror image by a continuous space transformation. The knot I deal with is the simplest after the trefoil knot, namely, the one which has only four double points under suitable projection. We shall see that the amphi-cheiral character of this knot corresponds to the more complicated nature of its automorphism group.

[**] See Math. Ann. 1910 and 1911 as well as the Münster dissertation of Gieseking. [The subscript $K\ell$ is for "Kleeblattschling", German for "trefoil knot" (Translator's note).]

Thus our topological problem leads to questions of auto-morphisms of an infinite group, and to the very general problem: given a presentation of a group by generators and defining relations, find a presentation of its automorphism group of the same type. In the case of the fundamental groups of closed surfaces we shall attack this problem in a work to appear shortly.

II.

The group $G_{K\ell}$ of the trefoil knot, as has been established several times[*], can be defined in the following way:

Generators:

$$a_1, a_2, a_3, a_4.$$

Relations :

$$a_1 a_4^{-1} a_2 = a_2 a_4^{-1} a_3 = a_3 a_4^{-1} a_1 = 1.$$

The associated group diagram consists of strips (see Fig. 4) which meet in threes at each boundary. If one represents the group diagram by a regular net in a non-euclidean space (see Math. Ann. 1911), the boundary of which is a cylinder with sides parallel to the strip edges, i.e. the generator a_4, then each element of the group corresponds to a certain motion of this geometry, which carries the group diagram, together with its

---

[*]See note [**] on the previous page.

labelling, into itself.   In particular, the generator a_4

resp. a_4^n corresponds to a rotation about an edge through

the origin by $\frac{2\pi}{3}$ resp. $\frac{2n\pi}{3}$ combined with a parallel displace-

ment along this edge through c resp. nc, where c is the

length of a_4.   In the plane perpendicular to the axis we

have the usual hyperbolic geometry.   If we intersect the

group diagram with such a plane through the origin 0, then the

cross-sections of the strips give an infinite line segment

complex Γ (see Fig. 5) : three segments emanate from each

point of Γ at angles of 120°.   Γ is a tree, i.e. there is

no closed polygon in Γ.   An element of $G_{K\ell}$ induces a motion

of Γ into itself.   <u>The generator a_4 and its powers are the</u>

<u>only elements which leave the origin 0 fixed</u>.

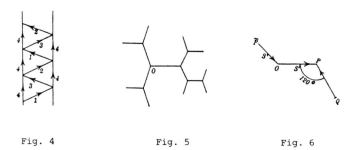

Fig. 4   Fig. 5   Fig. 6

1. Under an automorphism of $G_{K\ell}$ (i.e. an isomorphic mapping onto itself) a_4^3 goes into a_4^3 or a_4^{-3}. In fact, an element which commutes with all elements of $G_{K\ell}$ (a distinguished element of $G_{K\ell}$) must be sent to another such element. But a_4^3 and its powers are the only distinguished elements of G_K for which this holds, consequently the group $\{(a_4^3)^n\}$ goes into the group $\{(a_4^3)^n\}$ under the automorphism. But the group $\{(a_4^3)^n\}$ may be generated by one element, which may be chosen as either (a_4^3) or (a_4^{-3}), so a_4^3 must go into either a_4^3 or a_4^{-3}. However, we must still prove that $(a_4^3)^n$ are the only distinguished elements of $G_{K\ell}$.

Suppose now that S is any distinguished element, then in particular

$$Sa_4 S^{-1} = a_4$$

and thus causes a rotation about O. Now let O go into P, $\bar{P}$ into O, under the motion of Γ corresponding to S (see Fig. 6). Then under the motion Sa_4 the point O goes into P and the point $\bar{P}$, say, into Q, where $\sphericalangle\ OPQ = \frac{2\pi}{3}$ and $OP = PQ$. Consequently O goes into Q under the motion $Sa_4 S^{-1}$, because the line segment S, the projection S' of which on the plane of Γ is OP, is carried by the rotation a_4 about P into a segment with direction QP on the plane of Γ. Thus the element Sa_4S^{-1} only yields a rotation about 0 when P coincides with 0. Consequently S must be a power of a_4 to yield a rotation about 0, and all distinguished elements of G_K are of the form a_4^n. But

$$a_4 a_1 a_4^{-1} = a_2$$
$$a_4^2 a_1 a_4^{-2} = a_3$$

so neither a_4^{+1} or a_4^{+2} are distinguished elements, since a_1, a_2 and a_3 are different. On the other hand

$$a_4^3 a_1 a_4^{-3} = a_1$$
$$a_4^3 a_2 a_4^{-3} = a_2$$
$$a_4^3 a_3 a_4^{-3} = a_3$$

as the figure shows, from which it follows immediately that $(a_4^3)^n$ are the only distinguished elements, as was to be proved.

2. <u>Under an automorphism of $G_{K\ell}$, a_4 goes into Sa_4S^{-1} or $Sa_4^{-1}S^{-1}$.</u>

If a_4 goes into $\bar{a}_4$, then the third power of a_4 must be either a_4^3 or a_4^{-3}, by the above. Thus the motion which corresponds to $(\bar{a}_4)^3$ must leave Γ fixed. Thus $\bar{a}_4$ itself must cause a rotation of Γ of 120° about a real midpoint M of the plane of Γ. We now observe that Γ can be put together

from regular infinite polygons with angles of 120°, and with
vertices, after suitable choice of the length of side (or length
of the projection of a_1 etc.), lying on a limit circle (i.e.
a circle lying on the boundary cylinder). We shall call these
infinite polygons <u>fundamental polygons</u> for short. The fundamental polygons now go into one another by rotation about M.
If M lies in the interior of a fundamental polygon, then the
latter must go into itself by the rotation about M. For M
lies in the interior of only a single polygon, and M cannot
leave the interior of this polygon when it is rotated about M.
But the fundamental polygon cannot go into itself by a rotation
about a real point through an angle different from 2π. Furthermore, this is only possible when the midpoint for the rotation
coincides with the midpoint of the polygon, which is ideal in
any case (and lies on the boundary cylinder under a suitable
choice of segment length for Γ). Thus M cannot lie in the
interior of a polygon. Consequently M must lie on at least
one line segment of Γ. But if M lies on <u>one</u> segment, then
under rotation through 120° about M this goes into a different
segment which has only the point M in common with the original.
Thus M coincides with a vertex of Γ. Consequently $\bar{a}_4$
causes a rotation of 120° about a vertex M of Γ, and thus

yields a_4 or a_4^{-1} by conjugation with a suitable element S.

In fact, if S carries 0_1 into 0, and 0 into M (see Fig.7),

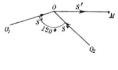

then 0 goes into itself by $S^{-1}\bar{a}_4 S$,

so $S^{-1}\bar{a}_4 S$ is a rotation about 0 and

hence $= a_4^n$. But n can only be $+1$

or -1, since $(\bar{a}_4)^3 = a_4^{+3}$ or a_4^{-3} and

$(\bar{a}_4)^3 = (Sa_4^n S^{-1})^3 = S^3 a_4^{3n} S^{-3} = a_4^{3n}$.

Fig. 7

3. If a_4 goes into a_4 under an automorphism, then a_1 goes into $a_4^{\alpha} a_1 a_4^{-\alpha}$.

The group $G_{K\ell}$ can be generated by a_1 and a_4 alone, for the other generators a_2 and a_3 are the conjugates of a_1 by a_4 and a_4^2 respectively. $G_{K\ell}$ is defined by <u>one</u> relation between a_4 and a_1, namely

$$a_4^{-2} a_1 a_4 a_1 = 1$$

which one obtains by elimination of a_2 and a_3 from the three original relations.

Now suppose we have an automorphism which sends a_4 into a_4, a_1 into $\bar{a}_1$. Then according to the relation we must have

$$\bar{a}_1 a_4 \bar{a}_1 = a_4^2$$

so that this is a rotation about 0. Now let (Fig. 8) OPP_1 be the projection of the polygonal path $\bar{a}_1\bar{a}_1$ on the plane of Γ. Then OPQ_1 is the projection of the polygonal path $\bar{a}_1 a_4 \bar{a}_1$ where

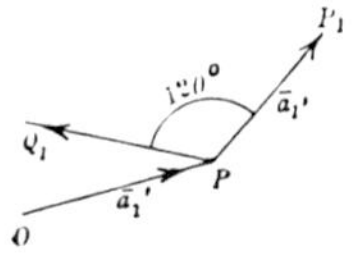

$$\sphericalangle \; P_1 P Q_1 \; = \; 120^{\circ}$$

and

$$PQ = PP_1 = OP.$$

Fig. 8

Accordingly $\bar{a}_1 a_4 \bar{a}_1$ can only be a_4^2, and thus a rotation about 0, when Q_1 coincides with 0, so that $\sphericalangle \; OPP_1 = 120^{\circ}$ (and in fact the rotation about P from 0 to P_1 must be right-handed when a_4 causes a left-hand rotation of 120° and conversely). We now use the projection $\bar{a}_1'$ of $\bar{a}_1$ to construct an infinite line segment complex $\bar{\Gamma}$ which has origin 0 and again three segments meeting at each vertex at angles of 120°. Since the length of $\bar{a}_1'$ is no less than the length of a_1' (the segment length of Γ), $\bar{\Gamma}$ is also a tree. It contains only vertices which are also vertices of Γ, but not all the vertices of Γ unless the length of $\bar{a}_1'$ equals that of a_1'. All vertices of the group diagram which can be obtained from a_4 and $\bar{a}_1$ project on to the vertices of $\bar{\Gamma}$, but since $\bar{a}_1$ and a_4 generate the whole group, $\bar{a}_1'$ must

have the segment length of Γ, and thus must equal
a_1^{+1}, a_2^{+1} or a_3^{+1}. But

$$a_4^{-2} a_i^{-1} a_4 a_i^{-1} \neq 1 \quad \text{for} \quad i = 1, 2, 3.$$

Thus $\bar{a}_1 = a_1$, a_2 or a_3, i.e. $= a_4^{\alpha} a_1 a_4^{-\alpha}$ by the above.
Since a_2 and a_3 are the conjugates of a_1 by a_4 and a_4^2
respectively, it follows that a_2 and a_3 are correspondingly
associated with $a_4^{\alpha} a_2 a_4^{-\alpha}$ and $a_4^{\alpha} a_3 a_4^{-\alpha}$.

4. It follows quite similarly that: <u>if a_4 goes into</u>
$\underline{a_4^{-1}}$ <u>then a_1 goes into $a_4^{\alpha} a_1^{-1} a_4^{-\alpha}$</u>. Then it follows from
the relation

$$a_4^{-2} a_1 a_4 a_1 = 1$$

that

$$a_4^{-1} a_1^{-1} a_4^2 a_1^{-1} = 1$$

(taking inverses), and by conjugation

$$a_4^3 a_4^{-1} a_1^{-1} a_4^2 a_1^{-1} a_4^{-3} = 1$$

so that, because a_4^3 commutes with all elements,

$$a_4^2 a_1^{-1} a_4^{-1} a_1^{-1} = 1.$$

Thus the same conclusions hold for a_4^{-1} and a_1^{-1} as we found under 3. for a_4 and a_2. <u>The mapping</u> $a_1 \mapsto a_1^{-1}$, $a_4 \mapsto a_4^{-1}$ <u>represents an outer automorphism of</u> $G_{K\ell}$. For a_1 is not conjugate to a_1^{-2}, as one sees most simply by making $G_{K\ell}$ abelian by addition of suitable relations. For then $G_{K\ell}$ goes into the ordinary infinite cyclic groups $\{a_1^{\,n}\}$, in which a_1 is not a conjugate of a_1^{-1}, otherwise $a_1^{\,2} = 1$ would follow.

5. If a_4 goes into Sa_4S^{-1} then there is an automorphism (namely the ordinary inner one) under which a_1 goes into Sa_1S^{-1}, and it follows from 4. that under the automorphism which sends a_4 into Sa_4S^{-1}, a_1 is sent into $a_4^{\alpha}Sa_1S^{-1}a_4^{-\alpha}$, for by introduction of the automorphism

$$a_4 \mapsto \overline{\overline{a}}_4 = Sa_4S^{-1}$$
$$a_1 \mapsto \overline{\overline{a}}_1 = Sa_1S^{-1}$$

we obtain an automorphism from the one given, in which $\overline{\overline{a}}_4$ goes into $\overline{a}_4$, and hence $\overline{a}_1$ must go into $a_4^{\alpha}\overline{\overline{a}}_1a_4^{-\alpha}$. Likewise a_1 goes into $Sa_4^{\alpha}a_1^{-1}a_4^{-\alpha}S^{-1}$ when a_4 goes into $Sa_4^{-1}S^{-1}$. But since by 2 there are no elements paired with a_4 other than the foregoing, we have the result:

<u>All automorphisms of</u> $G_{K\ell}$ <u>are given by the following correspondences:</u>

$$a_i \mapsto S a_i S^{-1}$$

and
$$a_i \mapsto S a_i^{-1} S^{-1} \qquad (i = 1,4)$$

respectively. Thus for a_2 and a_3 we obtain the correspondences

$$a_2 \mapsto S a_2 S^{-1}, \ a_3 \mapsto S a_3 S^{-1}$$

and
$$a_2 \mapsto S a_3^{-1} S^{-1}, \ a_3 \mapsto S a_2^{-1} S^{-1}.$$

<u>Using suitable inner automorphisms, all outer automorphisms can be brought into the form</u>

$$a_1 \mapsto a_1^{-1}, \ a_2 \mapsto a_3^{-1}, \ a_3 \mapsto a_2^{-1}, \ a_4 \mapsto a_4^{-1}.$$

6. Now if we have the two trefoil knots $K\ell_\lambda$ and $K\ell_\rho$, then the group of space curves relative to $K\ell_\lambda$ is isomorphically related to the group of space curves relative to $K\ell_\rho$ when one associates the generating curves a_1 and a_4 with the corresponding curves on the right in the opposite sense (see Fig. 9) (in which the latter curves result from those on the left by reflection in the plane of projection).

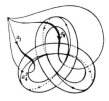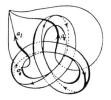

Fig. 9a   Fig. 9b

a_1 is a curve of latitude, $a_4^{\ 3}$ represents a curve of
longitude on both knots (see Fig.) (one results from the other
by reflection in the plane of projection). One sees
immediately that the indicatrix determined by a_1 and $a_4^{\ 3}$
in the case of $K\ell_\lambda$ is the opposite of that determined by
a_1 and $a_4^{\ 3}$ for $K\ell_\rho$. Thus this correspondence between the
curves of $K\ell_\lambda$ and $K\ell_\rho$ cannot correspond to a continuous
space transformation in which the curves go into each other.
But any other possible association of a_1 and a_4 with curves
of $K\ell_\rho$ corresponds to an automorphism of $G_{K\ell}$. We know that
all automorphisms of $G_{K\ell}$ may be reduced to the identity

$$a_1 \mapsto a_1$$
$$a_4 \mapsto a_4$$

or the correspondence

$$a_1 \mapsto a_1^{-1}$$
$$a_4 \mapsto a_4^{-1}$$

by means of an inner automorphism (transformation). As we
have seen, the first correspondence gives no possibility of
a continuous conversion of $K\ell_\lambda$ into $K\ell_\rho$, but neither does
the second, for a_1^{-1} and a_4^{-3} yield the same indicatrix
as a_1 and a_4^3.

Now to complete our proof we have only to show that
Sa_1S^{-1} and $Sa_4^3S^{-1}$ as latitude resp. longitude curves yield
the same indicatrix as a_1 and a_4^3. However, we can easily
show even more, namely, that the indicatrix is not altered
when one arbitrarily transforms (i.e. continuously deforms
in the complement space of U) the latitude curve a_1 and
the longitude curve a_4^3. In fact, in order to determine the
indicatrix it suffices to first determine the indicatrix given
by a_1 and the knot line with an assigned sense, say $\vec{K\ell}$, then
compare the sense of a_4^3 with that of $\vec{K\ell}$. But in the first
place the indicatrix with respect to $\vec{K\ell}$ and a_1 is not
altered by any continuous deformation of a_1 in the complement
space of $K\ell$, and in the second place a_4^3 cannot be continuously
deformed into a longitude curve of the opposite sense in relation
to $\vec{K\ell}$. We can give the latter the form $a_4^{-3}a_1^{12}$. This is

because all longitude curves can firstly be given the form $a_4^{\pm 3} a_1^{\,n}$, if necessary by transformation on the ring surface surrounding $K\ell$. Also, all these longitude curves transformable into each other on the ring surface have the same sense relative to $\vec{K\ell}$, and $a_4^{\,3} a_1^{\,n}$ has the opposite sense to $a_4^{-3} a_1^{\,n}$ in relation to $\vec{K\ell}$. But if $a_4^{\,3}$ can be transformed into $a_4^{-3} a_1^{\,n}$ in the complement space then n must equal 12, as follows immediately from the defining relations of $G_{K\ell}$ when one allows all elements to commute. It follows from

$$T a_4^{\,3} T^{-1} = a_4^{-3} a_1^{\,12}$$

that

$$a_4^{\,3} = a_4^{-3} a_1^{\,12}$$

because $a_4^{\,3}$ commutes with all elements. Thus we must have

$$a_4^{\,6} = a_1^{\,12}$$

a relation which is certainly not satisfied, as a glance at the group diagram shows. Thus the two curves which result from transformation of a_1 and $a_4^{\,3}$ yield the same indicatrix as a_1 and $a_4^{\,3}$, and the proof is complete.

III.

As we said in the beginning, we shall conclude by briefly treating an amphicheiral knot, i.e. a knot which is transformable into its own mirror image.* The group is obtained

---

*It was studied at my suggestion by a Kiel student. He has succeeded in obtaining the interesting group diagram for the knot.

directly by the process developed
in Math. Ann. 1910 by consideration of the four crossing
points where the five inner
regions meet:

Fig. 10.

Generators: a_1, a_2, a_3, a_4, a_5

Relations :

$$a_3 a_4^{-1} a_1 = a_1 a_2^{-1} a_3 = a_1 a_4^{-1} a_5 a_2^{-1} = a_3 a_2^{-1} a_5 a_4^{-1} = 1.$$

We consider the following correspondence between the
generators of the group G_L of our knot and the other elements:

$$a_1 \mapsto a_2 a_5^{-1} = b_1$$

$$a_2 \mapsto a_3 a_5^{-1} = b_2$$

$$a_3 \mapsto a_4 a_5^{-1} = b_3$$

$$a_4 \mapsto a_1 a_5^{-1} = b_4$$

$$a_5 \mapsto a_5^{-1} = b_5.$$

It is easy to see that this correspondence is an automorphism, in
which 1) all defining relations take the same form for the b_i
as for the a_i, 2) the a_i are expressible in terms of the b_i,
as follows immediately from the above relations.

A latitude curve β and a longitude curve λ for L are
represented by the elements:

$$a_1 \quad \text{resp.} \quad a_4 a_5^{-1} a_1 a_1 a_4^{-1} a_3$$

(λ can be best understood by proceeding from the point A
in the direction of the arrow). Under the isomorphism β
and λ go into

$$a_2 a_5^{-1} \quad \text{resp.} \quad a_1 a_2 a_5^{-1} a_2 a_3 a_5^{-1}.$$

The first curve again represents a latitude curve β'
which, together with the knot sensed as shown, $\vec{L}$, gives the
reverse indicatrix to β. The second curve is again a curve
of longitude, λ' (as one can see best by proceeding from
B in the direction of the arrow). Since λ' has the same
sense as λ relative to $\vec{L}$, the indicatrix (β, λ) is carried
into the reverse indicatrix (β', λ') by the automorphism.
This corresponds to the possibility of the knot being amphicheiral.
In fact, this isomorphism of G_L corresponds to a transformation
of L into its mirror image. One sees this easily when one
thinks of L spread over a spherical surface and converts the
outer region into the inner region 5. Then L goes into its
own mirror image (say, relative to the spherical surface).

It is natural to seek further automorphisms of G_L. One
finds a second one of the form:

$$c_1 = a_3^{-1} \quad, \quad c_2 = a_2^{-1}, \quad c_3 = a_1^{-1}, \quad c_4 = a_4^{-1}, \quad c_5 = a_5^{-1}.$$

This is also outer, like the first. If we compose these two automorphisms then we obtain, when we always associate an inner automorphism with the identity, a group of eight automorphisms which coincides with the eight-termed dihedral group. Whether this group yields all automorphisms by conjugation requires further investigation.

These automorphisms all have the property of carrying latitude and longitude curves into latitude and longitude curves again. Investigating whether this is the case for all automorphisms of knot groups will be an important step in mastering the general problem of transformability of knots into each other.

APPENDIX

ON THE GROUPS $A^a B^b = 1$ [1]   Otto Schreier

In what follows I derive a few properties of the groups $G_{a,b}$ generated by the elements A, B and defined by a relation $A^a B^b = 1$. The numbers a and b which may be assumed nonnegative, and which I assume to be > 1 in order to exclude trivial cases. These groups include the fundamental groups of all knots which can be embedded in the torus, e.g. $G_{3,2}$ is the group of the trefoil knot. Theorem II can be used to simplify

---

[1] I have presented these results at a topological seminar organized by Herr REIDEMEISTER in Vienna.

DEHN's proof of the distinctness of the two trefoil knots[2],
it avoids construction of the group diagram and applies word
for word to arbitrary knots on the torus[3].

I. **The numbers a,b are uniquely determined, up to order,
by the group $G_{a,b}$.**

We first construct the subgroup C of $G_{a,b}$ generated by
A^a $(= B^{-b})$. A^a is an invariant element* of $G_{a,b}$ therefore
C is a normal subgroup of $G_{a,b}$. Let $F = G_{a,b}/C$. F is
generated by the elements $A = CA$, $B = CB$, and defined by the
relations $A^a = 1$, $B^b = 1$. As is easy to see, each element of
F has a unique normal form

$$A^{x_1} B^{y_1} \ldots A^{x_p} B^{y_p}$$

where

$$0 \leqslant x_1 < a, \; 0 < x_2 < a, \; \ldots; \; \ldots \; 0 < y_{p-1} < b, \; 0 \leqslant y_p < b.$$

Lemma 1. F contains no invariant element other than 1.

Namely, if $A^{x_1} \ldots B^{y_p}$ is an invariant element in normal
form, then its commutativity with A and the uniqueness of the
normal form implies $y_p = 0$ and $p = 1$, similarly, commutativity
with B implies $x_1 = 0$, and thus our element is $A^0 B^0 = 1$.

---

[2] Math.Ann. 75 p.402.

[3] As Herr DEHN has pointed out to me, his proof is also appli-
cable to all these knots.

* Element which commutes with all others (Trans.).

Lemma 2. Each element of F of finite order is conjugate to a power of A or B.

Let Γ be an element of order c and let $A^{x_1} \ldots B^{y_p}$ be its normal form. Then at least one of the numbers $x_1, y_p = 0$, otherwise $\Gamma^c \neq 1$ because of the uniqueness of the normal form. This proves the assertion in case $p = 1$. Suppose it is true for all numbers $< p$. If say $y_p = 0$, then

$$A^{x_p} \Gamma A^{-x_p} = A^{x_1 - x_p} \ldots B^{y_{p-1}}$$

and the assertion holds for $A^{x_p} \Gamma A^{-x_p}$, hence also for Γ. (And analogously when $x_1 = 0$.)

It follows immediately from Lemma 1 that C is the centre of $G_{a,b}$. Lemma 2 now shows that $\max(a,b)$ is the greatest finite order for an element of the quotient of $G_{a,b}$ by its centre, and it is therefore uniquely determined by $G_{a,b}$. Now if we add commutativity of the generators to the defining relations of F then we obtain an abelian group of order ab; thus ab is the order of the abelianised quotient of $G_{a,b}$ by its centre, and thus uniquely determined by $G_{a,b}$. Thus Theorem I is proved.

In order to present the automorphism group of $G_{a,b}$, we prove

Lemma 3: When $a \neq b$ all the automorphisms of F are given by $\{A \mapsto T^{-1}A^r T, B \mapsto T^{-1}B^s T\}$ where r is prime to a and s is prime to b and T denotes an arbitrary element of F; when

$a = b$ there are the additional automorphisms

$$\{A \mapsto T^{-1} B^r T, \; B \mapsto T^{-1}A^s T\}.$$

The given substitutions clearly generate automorphisms of F. We have to show that there are no others. Now each automorphism must send A and B to elements A' and B' of orders a and b respectively. A' and B' cannot be conjugate to powers of the same generator of F, otherwise the other generator would have exponent sum 0 in each element generated by A', B' and thus would not appear in the group they generate. Thus when $a \neq b$ we must have $A' = P^{-1}A^r P$, $B' = \Sigma^{-1}B^s \Sigma$ where P, Σ are elements of F and r and s have the same meaning as above. If $\{A \mapsto A', \; B \mapsto B'\}$ determines an automorphism of F, so does $\{A \mapsto \Pi^{-1}A^r \Pi, \; B \mapsto B^s\}$ where $P\Sigma^{-1} = \Pi$. But since A must be generated from $\Pi^{-1}A^r \Pi$ and B^s, Π can only have the form $A^x B^y$. Now if we set $T = A^{-x}P = B^y\Sigma$, we have $A' = T^{-1}A^r T$, $B' = T^{-1} B^s T$. (Analogously for $a = b$.)

II. <u>All the automorphisms of</u> $G_{a,b}$ <u>for</u> $a \neq b$ <u>are given by</u> $\{A \mapsto T^{-1}A^\varepsilon T, \; B \mapsto T^{-1}B^\varepsilon T\}$ <u>where</u> $\varepsilon = \pm 1$ <u>and</u> T <u>is an arbitrary</u> <u>element of</u> $G_{a,b}$: <u>when</u> $a = b$ <u>one must add the automorphism</u> $\{A \mapsto T^{-1}B^\varepsilon T, \; B \mapsto T^{-1}A^\varepsilon T\}$.

Since C is a characteristic subgroup of $G_{a,b}$ it is mapped into itself by each automorphism, and each automorphism of $G_{a,b}$ uniquely determines automorphisms of C and F. If $\{A \mapsto A', \; B \mapsto B'\}$ determines an automorphism of $G_{a,b}$, then we must have $A'^a = A^{\varepsilon a} = B^{-\varepsilon b} = B'^{-b}$, and by Lemma 3, for $a \neq b$, $A' = T^{-1}A^{r+ha}T$, $B' = T^{-1}B^{s+kb}T$, where h, k are integers. It follows that $r + ha = \epsilon = s + kb$.   (Analogously for $a = b$.)

III. The automorphism group of $G_{a,b}$, for $a \neq b$, is generated by 3 elements I, J, K and defined by the relations $I^a = 1$, $J^b = 1$, $K^2 = 1$, $(KI)^2 = 1$, $(KJ)^2 = 1$ as an abstract group; for $a = b$ there is an additional generator L with the relations $L^2 = 1$, $KL = LK$, $LI = JL$.

In order to see this one has merely to denote the automorphisms $\{A \mapsto A, B \mapsto A^{-1}BA\}$, $\{A \mapsto B^{-1}AB, B \mapsto B\}$, $\{A \mapsto A^{-1}, B \mapsto B^{-1}\}$ of $G_{a,b}$ by I, J, K, and for $a = b$ denote the automorphism determined by $\{A \mapsto B, B \mapsto A\}$ by L; and to observe that C is the centre of $G_{a,b}$.

The next paper is an important unpublished work of Dehn :
notes of a lecture to the Breslau Mathematics Colloquium on
February 22, 1922.   The paper is mentioned a few times in
works of Baer and Goeritz (see below) but it exists only in the
form of a few mimeographed copies of handwritten notes, and
would probably have been lost if not for the careful preservation
of the Dehn Nachlass by Wilhelm Magnus.   The remaining mimeographed copies are mostly in the Humanities Library, Austin,
Texas.   In translating the notes I have also redrawn the
diagrams.

On the first two, introductory, pages there are some
interesting remarks on other lost proofs.   Dehn cites the
process of Poincaré [1904] (not 1905, as given by Dehn) for
deciding the number of topologically necessary intersections
of curves on a surface, and claims that the process can be
greatly simplified.   As I have mentioned in connection with
the simple curve problem (i.e., the special case of self-
intersection), this simplification may have been via Dehn's
algorithm for the conjugacy problem.   At any rate, Dehn
observes that such a process yields an algorithm for deciding
whether a given replacement of the generators, $a_k \mapsto a_k{}'$,
$b_k \mapsto b_k{}'$, in the fundamental group corresponds to a mapping class.
He is thereby assuming that any automorphism of the surface

group can be induced by a homeomorphism of the surface,
a result which he also claims, but for proof cites only a set
of lecture notes, Nielsen [1921].

A copy of this set has recently been found by Fenchel,
and my translation of it will appear in the forthcoming edition
of Nielsen's works.   Unfortunately, the proof is quite specific
to the genus 2 case, and the only clue it gives to the general
case is its use of the invariance of axis intersections
under automorphisms, a result which is proved generally in
Nielsen [1927, §9] and there credited in essence to Dehn.
A rather simple route from intersection invariance to a proof
that all automorphisms can be induced by homeomorphisms is
given in Marden [1976, p. 70].   After I spoke on the Dehn-
Nielsen theorem in Copenhagen in June 1983, Professor Borge
Jessen recovered a letter he had received from Nielsen, dated
11 December 1931, pointing out that the proof in Nielsen [1927]
can be replaced by a much simpler one.   Nielsen then sketches
a proof essentially the same as the one given 45 years later
by Marden.

Returning now to the paper before us: it is not very
clearly written, but one can see at least the following revolu-
tionary ideas : decomposition of the surface into 3-holed
spheres, determination of simple curve systems by intersection

and twist parameters at the boundaries of the 3-holed
spheres, i.e. by (6p-6)-dimensional integer vectors, where p
is the genus, and representation of mapping classes by transformations of this (6p-6)-dimensional space which, at least
in the special cases considered, are $\underline{\text{linear}}$[*] transformations.
Twist mappings appear, perhaps for the first time, and their
general significance is suggested.

Certain aspects of the curve representation and its
application to mappings were followed up in Baer [1927], [1928]
and Goeritz [1933a], [1933b], and the representation of mapping
classes was later investigated more thoroughly in Dehn [1938],
where it is shown that all mapping classes are represented by
linear transformations, generated by a finite number which
represent "twists". (For more on this, see the introduction
to the translation of Dehn [1938] below.) However, Dehn's
ideas did not begin to fulfil their potential until they were
rediscovered and given a new lease of life from the highly
original viewpoint of Thurston [1976] (see also Poenaru,
Laudenbach & Fathi [1979]). Part of the reason for Thurston's
success is that he reunites the combinatorial ideas with hyperbolic geometry, reversing Dehn's tendency to eliminate metric
proofs in favour of combinatorial ones.

---

[*] Though in certain cases the mapping breaks into different linear
pieces in cones with their apex at the origin (cf. p. 251).

REFERENCES

R. Baer [1927] : Kurventypen auf Flächen. _J. reine u. angew._

Math. 156, 231-246.

[1928] : Die Abbildungstypen der orientierbaren, geschlossenen

Flächen von Geschlecht 2. _J. reine u. angew. Math._ 160,

1-25.

M. Dehn [1938] : Die Gruppe der Abbildungsklassen.

Acta Math. 69, 135-206.

L. Goeritz [1933a] : Normalformen der Systeme einfacher Kurven

auf orientierbaren Flächen. _Abh. Math. Sem._

Univ. Hamburg 9, 223-243.

[1933b] : Die Abbildungen der Brezelflächen und Vollbrezel

vom Geschlecht 2. _Abh. Math. Sem. Univ. Hamburg_ 9,

244-259.

A. Marden [1976] : Isomorphisms between fuchsian groups. _Advances_

in Complex Function Theory (Eds. W. Kirwan, L. Zalcman).

Springer-Verlag, 56-78.

J. Nielsen [1921] : Die Abbildungstypen geschlossener Flächen und

ihre Beziehungen zu unendlichen Gruppen. (Notes of

Breslau lectures, 9 and 11 of March 1921).

[1927] : Untersuchungen zur Topologie der geschlossenen

zweiseitigen Flächen. _Acta Math._ 50, 189-358.

V. Poenaru, F. Laudenbach and A. Fathi [1979] : Travaux de Thurston

sur les surfaces. _Asterisque_, 66-67.

H. Poincaré [1904] : Cinquième complément à l'analysis situs.

Rend. circ. mat. Palermo 18, 45-110.

W. Thurston [1976] : On the geometry and dynamics of diffeomorphisms

of surfaces I (preprint).

ON CURVE SYSTEMS ON TWO-SIDED SURFACES,
WITH APPLICATION TO THE MAPPING PROBLEM

Lecture (supplemented) to the math. colloquium,
Breslau 11-2-1922   by Max Dehn

I Survey of previous results on the mapping problem: the fundamental
group of the closed surface has always been the starting point.

1) p = 1. Ordinary ring surface (torus)

Mapping class $S' = S^{\alpha_1} T^{\beta_1}$ where $\begin{vmatrix} \alpha_1 & \beta_1 \\ \alpha_2 & \beta_2 \end{vmatrix} = \pm 1$

$T' = S^{\alpha_2} T^{\beta_2}$

we therefore get the extended modular group.

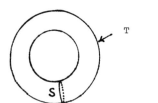

Fundamental group

$\begin{cases} \text{Generators S, T} \\ \text{Relation ST = TS} \end{cases}$

T

Connected with the period transformation of elliptic functions.

2) Two-sided surface with arbitrary p:

Fundamental group $\begin{cases} \text{Generators } a_1, \ldots, a_p, b_1, \ldots, b_p \\ \\ \text{Relation } \prod_{i=1}^{p} (a_i b_i a_i^{-1} b_i^{-1}) = 1 \end{cases}$

a) Necessary and sufficient conditions for the transformations

A $\begin{cases} a_k' = F_{1,k} (a_1, \ldots, a_p; b_1, \ldots, b_p) \\ \\ b_k' = F_{2,k} (a_1, \ldots, a_p; b_1, \ldots, b_p) \end{cases}$

to correspond to a mapping class are that, with fixed basepoint, a_k' and b_k' have no irreducible double points, a_k' and b_k' have <u>one</u> irreducible point of intersection, and otherwise no 2 of the 2p curves (with fixture of the basepoint) have an irreducible point of intersection. It can be decided whether the system (a_k', b_k') has this property following Poincaré (Pal. Rend. 1905), with the help of the diagram of the fundamental group in the hyperbolic plane. [The general process indicated by Poincaré can be greatly simplified.] Similarly, it can be decided by this method whether 2 mappings belong to the same class (i.e. whether one can be continuously carried into the other).

(b) Each <u>mapping class</u> corresponds to an <u>isomorphism</u> of the fundamental group with the same system A of transformations (immediately clear). Conversely: each isomorphism of the fundamental group corresponds to a mapping class with the same system t of transformations. (Dehn, see autogr. Vortrag of Nielsen, Breslau 9 and 11-3-21). Thus <u>the mapping problem</u> reduces to the isomorphism problem for the fundamental group. The proof uses essentially "topological" properties of the diagram of the fundamental group.

3) Two-sided double ring (p = 2). <u>The group of isomorphisms of the fundamental group is generated by 3 elements</u>. (Nielsen loc. cit.) The difficult proof depends on considerations from abstract group theory, essentially on the diagram of the fundamental group.

It turns out that the group of 4 x 4 matrices with determinant +1 is only a quotient of the group of isomorphisms. Only the positive value + 1 of the determinant comes into consideration here, because in a mapping which carries the fundamental curves into themselves, any reversal of orientation must occur in 2 fundamental curves simultaneously. A direct definition of the isomorphism group (without the detour of 2a) has not been possible as yet, because no means is at hand to present the relations between the generators of the isomorphism group.

II New approach (without use of the fundamental group)

The element of the fundamental group, i.e. the most general closed
oriented curve with a fixed initial point, proves to be too general in
one direction, and too special in the other, for the description of
topological relations on the surface of genus p > 1. The suitable object
of investigation for this purpose seems to be <u>the system of finitely many</u>
<u>curves without double points and intersection points, and not including</u>
<u>curves contractible to a point.</u> We shall consider two such systems to be
the same when one can be continuously deformed into the other on the
surface (i.e. when they belong to the same class).

1) Torus

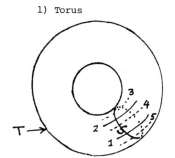

The curves of the system meet S in the
points 1, 2,..., m, which lie on S in
that order. We assume that no curve,
after leaving r, returns to S on the
same side at r + 1 or r - 1 (as can always

be arranged by continuous deformation). Then there is no curve at all which
returns to S on the same side from which it left. Thus the curves of the
system determine a permutation of the numbers 1,..., m, which because of the
absence of intersection points, is cyclic. To a number r there corresponds
the number r' at which the curve leaving S at r returns to S on the
(arbitrarily chosen) <u>positive</u> side. We have r' ≡ r + ν (mod m), where ν
is the total number of intersection points the curves of the system have with
T, each counted positively or negatively according as the curve crosses T
from the negative to the positive, or from the positive to the negative side

(the positive side is the side of T on which the point 1 lies).

This construction uniquely associates a number pair $\begin{pmatrix} m \\ \nu \end{pmatrix}$ with the curve system, as we have defined it, and conversely. (We have first assumed m > 0. We also set $\begin{pmatrix} m \\ \nu \end{pmatrix} = \begin{pmatrix} -m \\ -\nu \end{pmatrix}$ and associate $\begin{pmatrix} 0 \\ \nu \end{pmatrix}$ with the system of $|\nu|$ curves parallel to S. Then the scheme is complete.)

The number of closed curves of the system (all belonging to the same class) is equal to the number of cycles of the permutation $\begin{pmatrix} r \\ r + \nu \end{pmatrix}$ and thus equal to (m, ν). The transformations $\begin{pmatrix} m \\ \nu \end{pmatrix} \rightarrow \begin{pmatrix} \nu \\ m \end{pmatrix}$ and $\begin{pmatrix} m \\ \nu \end{pmatrix} \rightarrow \begin{pmatrix} m \\ \nu + m \end{pmatrix}$ (which do not change the number of cycles) correspond to mappings of the torus onto itself. Thus in particular, each pair $\begin{pmatrix} m \\ \nu \end{pmatrix}$, where (m, ν) = 1, can be reduced to $\begin{pmatrix} 1 \\ 0 \end{pmatrix}$ or $\begin{pmatrix} 0 \\ 1 \end{pmatrix}$. From this it follows that, using the mappings which correspond to these transformations, each simple closed curve U can be mapped onto the curve S. Under this mapping, any simple closed curve V which meets U once goes into a curve with the symbol $\begin{pmatrix} 1 \\ \nu \end{pmatrix}$, and under further transformations which leave S = $\begin{pmatrix} 0 \\ 1 \end{pmatrix}$ fixed, into $\begin{pmatrix} 1 \\ 0 \end{pmatrix}$, i.e. it is mapped onto T. We therefore have : any two simple curves which meet in one point can be mapped onto S and T by the mappings corresponding to the two symbol transformations. In order to separate the mapping classes completely from each other it is necessary to take account of orientation and to reflect this in the symbols. For this purpose we make the convention that when m > 0 the curves corresponding to $\begin{pmatrix} m \\ \nu \end{pmatrix}$ run towards the positive side of S, and also that when ν > 0 the curves corresponding to $\begin{pmatrix} 0 \\ \nu \end{pmatrix}$ have the same sense as S; finally, the curves corresponding to the symbol $\begin{pmatrix} -m \\ -\nu \end{pmatrix}$ shall have the opposite sense to those denoted by $\begin{pmatrix} m \\ \nu \end{pmatrix}$. This uniquely determines the orientation in all cases.

From $\begin{pmatrix} m \\ \nu \end{pmatrix}$ we obtain the following by successive transformations:

$$\begin{pmatrix} m \\ \nu \end{pmatrix} \rightarrow \begin{pmatrix} \nu \\ m \end{pmatrix} \rightarrow \begin{pmatrix} \nu \\ m - \nu \end{pmatrix} \rightarrow \begin{pmatrix} m - \nu \\ \nu \end{pmatrix} \rightarrow \begin{pmatrix} m - \nu \\ m \end{pmatrix} \rightarrow \begin{pmatrix} m \\ m - \nu \end{pmatrix} \rightarrow \begin{pmatrix} m \\ -\nu \end{pmatrix}$$

Similarly:

$$\begin{pmatrix} m \\ \nu \end{pmatrix} \rightarrow \begin{pmatrix} \nu \\ m \end{pmatrix} \rightarrow \begin{pmatrix} \nu \\ -m \end{pmatrix} \rightarrow \begin{pmatrix} -m \\ \nu \end{pmatrix}$$

Consequently, each pair $\begin{pmatrix} m_1 \\ \nu_1 \end{pmatrix}$ and $\begin{pmatrix} m_2 \\ \nu_2 \end{pmatrix}$ of symbols which represent

2 simple curves U and V with a single point of intersection can be carried

into the 4 pairs of symbols $\begin{pmatrix} 0 \\ \pm 1 \end{pmatrix}$ and $\begin{pmatrix} \pm 1 \\ 0 \end{pmatrix}$. Each can be obtained by a mapping,

using the mappings which correspond to the transformations. Thus some already

known results are derived in a new way.

   2) <u>Determination of a curve system on the surface of genus p by</u>

<u>3p - 3 pairs of numbers.</u>

       a) p = 2

upper half

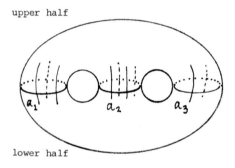

The system has m_i intersections

with a_i. By continuous deformation

we can always arrange that no curve

which leaves a_i at r returns at

r ± 1 so that the intervening piece

lower half   of curve together with the piece between

r and r ± 1 on a_i is contractible to a point. The $m_1 + m_2 + m_3$ curve pieces

unite in pairs on the "upper and "lower halves of the double ring ($m_1 + m_2 + m_3$

must therefore be even). Such a half is a sphere with 3 holes a_i. If we extend

the holes along the pieces of the curve system connecting them, until the

latter are replaced by direct contact of the a_i, then, taking all properties

of the curve system into consideration, we obtain only the following possibilities

of combination, corresponding to the 3 topologically different ways of covering

a sphere by 3 discs.

Case a) $m_i + m_k > m_\ell$

    Here we let μ_1 be the number of pieces leaving a_2 which unite on either

half with pieces coming from a_3. Analogously for μ_2 and μ_3

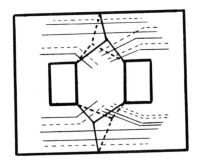

$$\mu_i = \frac{m_k + m_\ell - m_i}{2}$$

The corresponding "seam diagram"

is

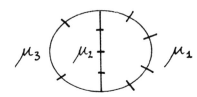

Case b) $m_i + m_k < m_\ell$, say m_2

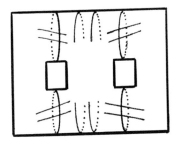

Seam diagram

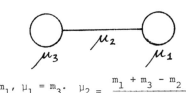

$$\mu_3 = m_1, \quad \mu_1 = m_3. \quad \mu_2 = \frac{m_1 + m_3 - m_2}{2}$$

$\mu_1 (\mu_3)$ is the number of pieces emanating from a_2 which unite in each half with pieces from a_3 (a_1) – μ_2 is the number of pieces emanating from a_2 which unite in each half with the others from a_2.

Case a, b) $m_i + m_k = m_\ell$, say $= m_2$

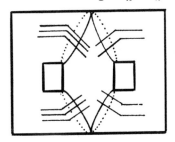

Seam diagram

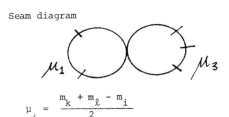

$$\mu_i = \frac{m_k + m_\ell - m_i}{2}$$

The μ_i and hence the seam diagram are thus completely determined by the numbers m_i. For this reason the lower half always has the same seam diagram as the upper half. On each of the seam diagrams there are $\frac{m_1 + m_2 + m_3}{2}$ points, which lie above and below in the same way on the 3 (resp. 2) segments of the seam diagram. We can speak of corresponding points on the two figures. We shall give each point a double designation, corresponding to the two pieces which unite in this point, say 1, 1 to 1, m_1; 2, 1 to 2, m_2; 3, 1 to 3, m_3. The point with the designation $i,r = k,s$ is where piece number r of the ith division unites with piece number s of the kth division. The pieces themselves are numbered in the same way as their intersection points with the a_i. If we leave one of the m_i pieces of the ith division from the upper seam diagram, then we again return from the lower to the upper figure, but in general not at the same place, because the piece emanating from the point i,r does not meet the lower figure at the point i,r but at the point i,r', where $r' \equiv r + \nu \pmod{m_i}$.

Thus the curve system generates a substitution of the numbers of the upper seam diagram (the numbers r_i are determined up to multiples of m_i by the remainder). In order to fix the ν_i geometrically, we start with the simplest case, in which the same 2 curve pieces are joined together on the lower seam figure as on the upper. The system then consists of $\dfrac{m_1 + m_2 + m_3}{2}$ closed curves.

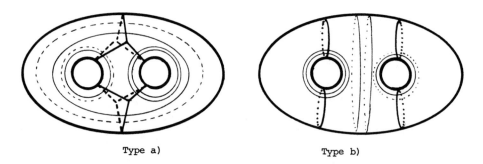

Type a)   Type b)

The simplest case also has the property that for type a) the system has no intersection points with the curves b_i (see fig.), i.e. the curves lie wholly on the front or wholly on the back of the surface, and that for type b) the system has only $m_\ell - m_i - m_k$ such intersection points with the curve b_ℓ

We now fix the points on the seam figure and construct the most general system by "twisting" the m_i curve pieces of the 3 divisions. The number $|\nu_i|$ is then the number of intersection points of the m_i curve pieces with b_k as well as with b_ℓ. The sign of ν_i is taken to be positive, say, when the m_i curve pieces of the ith division constitute a right-handed screw system. We denote the curve system by $\begin{pmatrix} m_1 & m_2 & m_3 \\ \nu_1 & \nu_2 & \nu_3 \end{pmatrix}$. One can also follow the course of the curves of the system on the seam diagram by inscribed polygons with the displacement segments ν_i (cf. in the torus case the role of the m-gon within the m-gon, with the displacement segment ν).

Example $\begin{pmatrix} 4 & 5 & 3 \\ 2 & 3 & 2 \end{pmatrix}$ $\mu_1 = 2$, $\mu_2 = 1$, $\mu_3 = 3$

Divides into 2 closed curves

Seam diagram

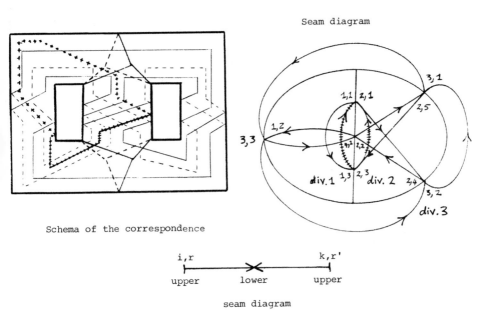

Schema of the correspondence

seam diagram

The two substitutions for the upper seam diagram:

(1,3)   (1,1 3,1 2,2 2,4 3,3)

(2,3)   (2,1 1,2 3,2 1,4 2,5)

The 2nd substitution is always the inverse of the first, when symbols for coincident points are regarded as the same. The number of cycles of the substitution equals the number of closed curves of the system. Each curve system on the double ring (of simple, disjoint, non-contractible curves) is uniquely determined by 3 pairs of numbers. The 3 second terms ν_1, ν_2, ν_3 are arbitrary, but the first terms of the pairs must be $\geqslant 0$, and their sum must be even. We shall understand $\begin{pmatrix} 0 & m_2 & m_3 \\ \nu_1 & \nu_2 & \nu_3 \end{pmatrix}$ to be the system that consists of $\begin{pmatrix} 0 & m_2 & m_3 \\ 0 & \nu_2 & \nu_3 \end{pmatrix}$ and ν_1 curves parallel to a_1.

b) p > 2

m_i intersection points of the curve system with a_i (under the assumption, which is always achievable, that no curve bounds a disc in conjunction with a

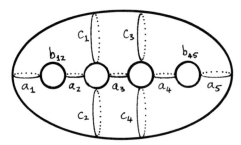

piece of a_i). Σm_i must be even. For p > 2 it is no longer convenient, as it was with p = 2, to depend upon the seam diagrams of the upper and lower halves. As one quickly realizes, even for p = 3, there are 5 different types of seam diagram, and the type is by no means uniquely determined by the m_i, as was the case for p = 2. Moreover, for each system of m_i there are infinitely many seam diagrams not continuously deformable into each other on the surface. However, the following construction enables the case p > 2 to be carried back to the case p = 2 by a kind of reduction.

Suppose, say, that $p = 4$. Then in the upper and lower halves we add

one of the curves c_1 resp. c_2 meeting the curve b_{15} and b_{23} each

once, similarly on both halves, one of the curves c_3 resp. c_4 meeting

of the curves b_{15} and b_{34} each once (see fig.) Then the upper and

lower halves both divide into 3 spheres, each with 3 holes. Let ℓ_i be

the number of points of intersection of the curves of the system with c_i,

(again under the hypothesis that no curve bounds a disc in conjunction with

a piece of c_i). Then by II, 2a the kind of connection on each one of the

spheres of the 6 spheres is uniquely determined by the m_i and b_i. From any

(normal) system with given m_i and ℓ_i we obtain the most general one with the

same numbers by twists through the "angle" $\frac{2\pi}{m_i} \nu_i$ resp. $\frac{2\pi}{\ell_i} \delta_i$ along the

boundary lines a_i and c_i between the spheres. A convenient way to

construct the normal system is to begin with one of the 3-holed spheres and

then construct normal systems of connections for all the spheres in a fixed

order, following the construction of the normal system (simplest case) for

$p = 2$. Thus the most general curve system on the surface of genus $p = 4$

is uniquely determined by $5 + 4$ number pairs. The second terms of these

number pairs are any integers; the first terms are all ≥ 0 and satisfy the

following conditions:

$$m_1 + m_2 + \ell_1, \ m_1 + m_2 + \ell_2, \ m_3 + \ell_1 + \ell_3, \ m_3 + \ell_2 + \ell_4, \ \ell_3 + m_4 + m_5,$$

$$\ell_4 + m_4 + m_5 \qquad\qquad \text{are even}$$

In general we have:

The most general system of disjoint, simple, non-contractible curves

on a closed surface of genus p is uniquely determined by $3p - 3$ number

pairs

$$\begin{pmatrix} m_1 & m_2 & \ell_1 & \ell_2 & m_3 & \ell_3 & \ell_4 & m_4 & \cdots & m_{p-1} & \ell_{2p-5} & \ell_{2p-4} & m_p & m_{p+1} \\ \nu_1 & \nu_2 & \delta_1 & \delta_2 & \nu_3 & \delta_3 & \delta_4 & \nu_4 & \cdots & \nu_{p-1} & \delta_{2p-5} & \delta_{2p-4} & \nu_p & \nu_{p+1} \end{pmatrix}$$

The numbers $\quad m_1 + m_2 + \ell_1,\ \ m_1 + m_2 + \ell_2,\ \ldots$

$$\ldots\ \ell_{2k-5} + m_k + \ell_{2k-3},\ \ \ell_{2k-4} + m_k + \ell_{2k-2}\ \ \ldots$$
$$\ldots\ \ell_{2p-5} + m_p + m_{p+1},\ \ \ell_{2p-4} + m_p + m_{p+1}\qquad \text{are even}$$

The numbers in the upper row are cardinalities, namely the numbers of intersection points of the curves of the given system with the $3p-3$ curves $a_i, \ldots, a_{p+1},\ c_1, \ldots, c_{2p-4},$ by means of which the surface is divided into $2(p-1)$ three-holed spheres. The numbers in the lower row are arbitrary positive or negative integers and they denote the twists which convert a normal system with the same upper row into the given system.

Since for $p = 1$ the class is determined by <u>one</u> pair, and for $p = 0$ no distinct classes exist at all, we have an exact analogue to Riemann's theorem on classes of algebraic functions of the same genus. (Connection?)

3) Transformation, reduction of systems of number pairs

The association of number pairs with curve systems is important first of all because it is related in a simple, natural way to the course of the curves in the system, so that one may expect that significant geometric properties of the system are expressed in simple arithmetic properties of the system of number pairs. This expectation is fulfilled.

a) $p = 2$

We consider the following 4 continuous mappings of the surface onto itself.

I Rotation about the triple symmetry axis by $\dfrac{2\pi}{3}$

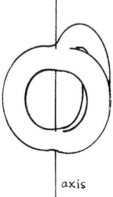

II Reflection in the equatorial plane (through $b_1,\ b_2,\ b_3$).

axis

III Exchange of the outer and inner spaces (or upper and lower, back and front)

IV Twisting the tube neighbouring a_1 through 2π

I, II and III extend to continuous mappings of the whole space onto itself, and therefore carry curve systems into ones which are isotopic relative to all of space

I corresponds to cyclic interchange $\begin{pmatrix} m_i & \nu_i \\ m_{i+1} & \nu_{i+1} \end{pmatrix}$

II corresponds to $\begin{pmatrix} \nu_i \\ -\nu_1 \end{pmatrix}$

IV corresponds to $\begin{pmatrix} \nu_1 \\ \nu_1 + m_1 \end{pmatrix}$

III is somewhat more difficult to express arithmetically.

For type a) $(m_i + m_k \geqslant m_\ell)$ and $\nu_i \geqslant 0$, III corresponds to tne arithmetic transformation: $\begin{pmatrix} m_i & \nu_i \\ \nu_k + \nu_\ell & m_k + m_\ell \dfrac{\cdot}{2} m_1 \end{pmatrix}$

Example $\begin{pmatrix} 4 & 5 & 3 \\ 2 & 3 & 2 \end{pmatrix} \xrightarrow{\text{IV}^{-1},\ \text{I}} \begin{pmatrix} 4 & 5 & 3 \\ -2 & -2 & -1 \end{pmatrix}$

$\xrightarrow{\text{II}} \begin{pmatrix} 4 & 5 & 3 \\ 2 & 2 & 1 \end{pmatrix}$

$\xrightarrow{\text{III}} \begin{pmatrix} 3 & 3 & 4 \\ 2 & 1 & 3 \end{pmatrix}$

$\xrightarrow{\text{I, IV}^{-1},\text{II}} \begin{pmatrix} 3 & 3 & 4 \\ 1 & 2 & 1 \end{pmatrix}$

$\xrightarrow{\text{III}} \begin{pmatrix} 3 & 2 & 3 \\ 2 & 2 & 1 \end{pmatrix}$

$\xrightarrow{\text{IV}^{-1},\text{II}} \begin{pmatrix} 3 & 2 & 3 \\ 2 & 0 & 1 \end{pmatrix}$

$\xrightarrow{\text{III}} \begin{pmatrix} 1 & 3 & 2 \\ 1 & 2 & 1 \end{pmatrix}$

$\xrightarrow{\text{I,IV}^{-1},\text{II}} \begin{pmatrix} 1 & 3 & 2 \\ 0 & 1 & 1 \end{pmatrix}$

$\xrightarrow{\text{III}} \begin{pmatrix} 2 & 1 & 1 \\ 2 & 0 & 1 \end{pmatrix}$

$\xrightarrow{\text{I,IV}^{-1}} \begin{pmatrix} 2 & 1 & 1 \\ 0 & 0 & 0 \end{pmatrix}$

$\xrightarrow{\text{III}} \begin{pmatrix} 0 & 0 & 0 \\ 0 & 1 & 1 \end{pmatrix}$

and thus the system $\begin{pmatrix} 4 & 5 & 3 \\ 2 & 3 & 2 \end{pmatrix}$ consists of two non-separating curves not belonging to the same class. (cf. fig. p. 242)

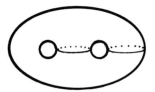

The general arithmetic form of the transformation III will be given here only for the m_i, which is sufficient for our purpose:

$$m_i \mapsto \frac{|v_k + v_\ell| + |v_k| + |v_\ell|}{2} + \left| \frac{|v_k| + |v_\ell| - |v_k + v_\ell|}{2} - \mu_i \right| - \mu_i$$

where we set

$$\mu_i = m_k + m_\ell - \frac{|m_i + m_\ell - m_k| + |m_i + m_k - m_\ell|}{2}$$

The transformations carry the types into each other. One sees without difficulty that each number pair triple can be converted, by means of the 4 transformations, into a triple in which each pair contains at least <u>one</u> zero, and indeed into the following triples:

1) $\begin{pmatrix} 0 & 0 & 0 \\ v_1 & v_2 & v_3 \end{pmatrix}$

2) $\begin{pmatrix} 0 & m_2 & 0 \\ v_1 & 0 & v_3 \end{pmatrix}$

3) $\begin{pmatrix} 0 & \mu_2 + m_3 & m_3 \\ v_1 & 0 & 0 \end{pmatrix}$

4) $\begin{pmatrix} m_1 & \mu_2 + m_1 & 0 \\ 0 & 0 & v_3 \end{pmatrix}$

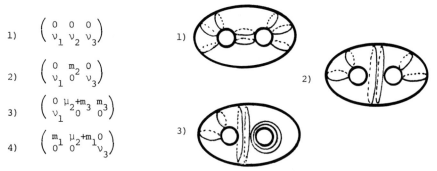

To prove reducibility it suffices to snow that, by suitable transformations, the maximum of the m_i can always be reduced, until one of the 3 minimal systems above is reached. Now let $\begin{pmatrix} m_1 & m_2 & m_3 \\ v_1 & v_2 & v_3 \end{pmatrix}$ be the given system, $m_1 \geqslant m_2 \geqslant m_3$, then, if we do not have $m_1 = m_2 = m_3$,

m_1 is transformed by III into $m_1' < m_1$. Namely, for type a)

$m_1' = \dfrac{m_2 + m_3}{2} < m_1$, for type b) $m_1' = \dfrac{m_2 + m_3}{2} + m_1 - m_2 - m_3 < m_1$.

Then one uses II and IV to make $|v_1| \leqslant \dfrac{m_1}{2}$, then $m_2' \leqslant \dfrac{m_1}{2} + \dfrac{m_2}{2} \leqslant m_1$,

$m_3' \leqslant \dfrac{m_1}{2} + \dfrac{m_3}{2} < m_1$, so that fewer numbers attain maximal value m_1 after

the transformation than before. Only the case $m_1 = m_2 = m_3$ remains. In

case we do not have $0 < v_1 = \dfrac{m_i}{2}$, one can begin as before by reducing one

of the m_i, without making the others larger. However, if $v_i = \dfrac{m_i}{2}$ one

uses II to replace v_1 by $-v_1$, and then obtains the desired reduction

similarly.

Example: $\begin{pmatrix} 3 & 20 & 3 \\ 1 & -10 & 1 \end{pmatrix} \rightarrow \begin{pmatrix} 9 & 16 & 9 \\ 10 & -7 & 10 \end{pmatrix} \rightarrow \left[\begin{pmatrix} 9 & 16 & 9 \\ 1 & -7 & 1 \end{pmatrix} \right]$

$\rightarrow \begin{pmatrix} 6 & 2 & 6 \\ 8 & -3 & 8 \end{pmatrix} \rightarrow \left[\begin{pmatrix} 6 & 2 & 6 \\ 2 & 1 & 2 \end{pmatrix} \rightarrow \right] \rightarrow \begin{pmatrix} 3 & 4 & 3 \\ 1 & 5 & 1 \end{pmatrix}$

$\rightarrow \left[\begin{pmatrix} 3 & 4 & 3 \\ 1 & 1 & 1 \end{pmatrix} \rightarrow \right] \rightarrow \begin{pmatrix} 2 & 2 & 2 \\ 2 & 1 & 2 \end{pmatrix} \rightarrow \left[\begin{pmatrix} 2 & 2 & 2 \\ 0 & 1 & 0 \end{pmatrix} \rightarrow \right]$

$\rightarrow \begin{pmatrix} 1 & 0 & 1 \\ 1 & 1 & 1 \end{pmatrix} \rightarrow \left[\begin{pmatrix} 1 & 0 & 1 \\ 0 & 1 & 0 \end{pmatrix} \rightarrow \right] \rightarrow \begin{pmatrix} 0 & 1 & 1 \\ 1 & 0 & 0 \end{pmatrix}$

$=$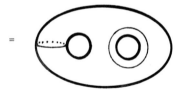

b) $p = 3$

Normal system

We base our discussion on a decomposition of the $p = 3$ surface

into 4 three-holed spheres, so placed as to exhibit tetrahedral symmetry

(see fig.) Correspondingly, we denote curves denoted a_i and c_i

above by a_1 to a_6, and the numbers m_i and ℓ_i by $m_1, \ldots, m_6$, and

the numbers v_i and δ_i by $v_1, \ldots, v_6$. To determine a normal system

we shall associate each of the curves a_i with one of the four curves

b_I, b_{II}, b_{IV}, b_V intersecting it in the external space, say:

$$a_1 - b_I,\ a_2 - b_V,\ a_3 - b_{II},\ \left.\begin{matrix} a_4 \\ a_5 \\ a_6 \end{matrix}\right\} - b_{IV}$$

For the sake of simplicity we think of all the intersection points of the curves lying on the front side of the surface. We can now place the connecting segments in such a way that none of them are intersected by the curves b_I, b_{II}, b_{IV}, b_V except those which return to the same a_i from which they depart. These will be placed in such a way that they have an intersection point with the a_i on the piece of the sphere in question lying opposite b_7, and an intersection point with the b_k associated with the a_i, cf. the example shown,

$$\begin{pmatrix} 3 & 2 & 3 & 1 & 0 & 1 \\ 0 & 0 & 0 & 0 & 0 & 0 \end{pmatrix}$$

This determines a curve system with the numbers m_i uniquely up to continuous deformation. If we take, say, the right handed screw system as positive, then it becomes uniquely determined which system is represented by the symbol $\begin{pmatrix} m_1 & m_2 & m_3 & m_4 & m_5 & m_6 \\ v_1 & v_2 & v_3 & v_4 & v_5 & v_6 \end{pmatrix}$

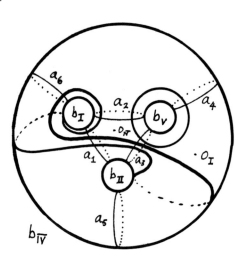

Transformations

I a) Rotation through $\frac{2\pi}{3}$ about the 3-fold symmetry axis through the "midpoint" 0_{IV} of the "triangle" 1 3 3 = V II I

b) Rotation through $\frac{2\pi}{3}$ about the 3-fold symmetry axis through the "midpoint" 0_I of the "triangle" 3 4 5 = IV II V

II Reflection in the equatorial plane through I, II, IV, V.

III In order to fix a particular kind of exchange of internal and
external space, we introduce (see fig.) the curve b_{II} on the front

side enclosing b_I and b_{II}, and the
curve b_{VI} on the back side,
enclosing b_I and b_V. Then a
mapping of the surface which
exchanges the internal and
external spaces is determined
by the following exchange rules
between the curves a_i and b_k:

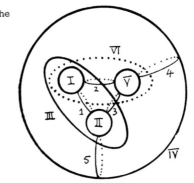

$1 \leftrightarrow$ I (with orientation), $2 \leftrightarrow$ II, $3 \leftrightarrow$ III, $4 \leftrightarrow$ IV, $5 \leftrightarrow$ V, $6 \leftrightarrow$ VI.

This is our transformation III.

IV Twist of the neighbourhood of a_1 through 2π.

Ia corresponds to the substitution

$$\begin{pmatrix} m_1 & m_2 & m_3 & m_4 & m_5 & m_6 \\ m_2 & m_3 & m_1 & m_5 & m_6 & m_4 \end{pmatrix}$$

and

$$\begin{pmatrix} \nu_1 & \nu_2 & \nu_3 & \nu_4 & \nu_5 & \nu_6 \\ \nu_2 & \nu_3 & \nu_1 & \nu_5 & \nu_6 & \nu_4 \end{pmatrix}$$

b " " " "

$$\begin{pmatrix} m_1 & m_2 & m_3 & m_4 & m_5 & m_6 \\ m_2 & m_6 & m_4 & m_5 & m_3 & m_1 \end{pmatrix}$$

and

$$\begin{pmatrix} \nu_1 & \nu_2 & \nu_3 & \nu_4 & \nu_5 & \nu_6 \\ \nu_2 & \nu_6 & \nu_4 & \nu_5 & \nu_3 & \nu_1 \end{pmatrix}$$

II corresponds to $\begin{pmatrix} \nu_i \\ -\nu_i \end{pmatrix}$

IV corresponds to $\begin{pmatrix} \nu_1 \\ \nu_1 + m_1 \end{pmatrix}$

III, as for the case $p = 2$, is complicated, and will not be investigated
further here.

Whether the 5 transformations I-IV suffice for a complete reduction,
as do the 4 transformations I-IV in the case $p=2$, is doubtful.

4. Continuous (homeomorphic) mappings of the $p = 2$ surface onto
itself.

The minimal systems attainable by the transformations are in
part "equivalent" to each other, i.e. convertible into each other by
continuous surface mappings. (See Enzyklopädie, Analysis situs B7),
and indeed obviously

$$\begin{pmatrix} 0 & 0 & 0 \\ v_1 & v_2 & v_3 \end{pmatrix} \text{ is equivalent to } \begin{pmatrix} 0 & 0 & 0 \\ v_3 & v_2 & v_1 \end{pmatrix}, \text{ likewise}$$

$$\begin{pmatrix} 0 & m_2 & 0 \\ v_1 & 0 & v_3 \end{pmatrix} \quad \text{to} \quad \begin{pmatrix} 0 & m_2 & 0 \\ v_3 & 0 & v_1 \end{pmatrix}, \text{ and}$$

$$\begin{pmatrix} 0 & \mu_2{+}m_3 & m_3 \\ v_1 & 0 & 0 \end{pmatrix} \quad \text{to} \quad \begin{pmatrix} m_3 & \mu_2{+}m_3 & 0 \\ 0 & 0 & v_1 \end{pmatrix},$$

for they are mapped into each other when the left and right halves of
the double ring are exchanged. We add these transformations, as V, to
those previously given for $p = 2$. Further, it is easy to see that

$$\begin{pmatrix} 0 & 2\mu_2{+}m_3 & m_3 \\ v_1 & 0 & 0 \end{pmatrix} \text{ is equivalent to } \begin{pmatrix} 0 & 2\mu_2 & 0 \\ v_1 & 0^2 & m_3 \end{pmatrix} - \text{ it is exchange of latitude}$$

and longitude in one (the right) half of the double ring, whereby the
v_1 curves parallel to a_1 and the μ_2 separating curves go into each
other, and the m_3 curves parallel to b_1 go into m_3 curves parallel
to a_3. We add such a transformation, as VI, to the above, and note that
we can construct the transformation III by combining VI and V. Thus it
follows that :

I, II, IV, V, VI generate each class of continuous mappings of the
$p = 2$ surface onto itself. For with these 5 transformations we can first
transform a system of 3 curves U_1, U_2, W, where W is separating and
non-contractible, U_1 and U_2 non-separating, and

disjoint from W and each other, into the system $(a_1, a_3, d_2) = \begin{pmatrix} 0 & 2 & 0 \\ 1 & 0 & 1 \end{pmatrix}$, where $d_2 = \begin{pmatrix} 0 & 2 & 0 \\ 0 & 0 & 0 \end{pmatrix}$

As a result, any 2 curves which meet U_1, U_2 in single points but do not meet W, go into curves $\begin{pmatrix} 1 & 1 & 0 \\ \nu_1 & \nu_2 & 0 \end{pmatrix}$ and $\begin{pmatrix} 0 & 1 & 1 \\ 0 & \nu_2 & \nu_3 \end{pmatrix}$, and by further mappings corresponding to IV and I, into $\begin{pmatrix} 1 & 1 & 0 \\ 0 & 0 & 0 \end{pmatrix}$ and $\begin{pmatrix} 0 & 1 & 1 \\ 0 & 0 & 0 \end{pmatrix}$, i.e. into b_3 and b_1. But our mappings, namely II (reflection in the vertical plane through the b's) transformed with the mapping III, also enable a_1 and b_3 to be mapped into themselves with reversal of orientation, where d_2, a_3 and b_1 go into themselves (by the mapping II VI $(II)^{-1}$ $(VI)^{-1}$, which sends a_1 and b_3 into themselves, a_3 and b_1 into a_3^{-1} and b_1^{-1} resp.) Thus we have shown that each mapping can be composed from these 5 mappings, for we have already remarked in the introduction that, when the orientation of a_1, a_3 and b_3 is given, the orientation of b_1 is also determined. But a mapping which carries a_1, a_3, b_1, b_3 into themselves without change of orientation belongs to the identity class.

This derives the result of Nielsen in a direct, elementary way. That 5 generators were used here matters little, for naturally one can show that the transformations I, II, IV, V, VI can be composed from the 3 Nielsen transformations. We have also seen that these transformations can be represented by linear transformations of the 6 dimensional (μ_i, ν_i) number lattice, and indeed those corresponding to I, II, IV, VI are the same for all lattice points. On the other hand V is different in different sectors with the zero of the lattice at the vertex. This holds incidentally, for the 5 transformations I-IV for p=3.

Thus the way is opened for a precise study of the mapping class group for p=2. Moreover, it is to be expected that the solution of the problem for p>2 will involve no further serious difficulties. It should be noted that the peculiarly non-analytic character of the set of linear transformations is not to be avoided. On the contrary, a greater symmetry is perhaps attainable by introduction of a second seam diagram, corresponding to the transformation III.

5. Knots and links

Our representation brings the gain to knot and link theory that one is able to represent the general knot or link uniquely by a series of number pairs. Further: the transformations I, II, III for p=2 and p=3 transform knots and links on the surface into others which are spatially equivalent (isotopic). The same holds for the special case of IV.

$$\begin{pmatrix} 1 & m_2 & m_3 \\ \nu_1 & \nu_2 & \nu_3 \end{pmatrix} = \begin{pmatrix} 1 & m_2 & m_3 \\ 0 & \nu_2 & \nu_3 \end{pmatrix} \text{ resp. } \begin{pmatrix} 1 & m_2 & m_3 & m_4 & m_5 & m_6 \\ \nu_1 & \nu_2 & \nu_3 & \nu_4 & \nu_5 & \nu_6 \end{pmatrix} = \begin{pmatrix} 1 & m_2 & m_3 & m_4 & m_5 & m_6 \\ 0 & \nu_2 & \nu_3 & \nu_4 & \nu_5 & \nu_6 \end{pmatrix}$$

Thus for the first time we have the possibility of systematically treating previously unassailable problems of the reduction of knots and links. However, the 4 transformations certainly do not suffice for a complete explanation in the case p=2 or p=3.

Example. Trefoil knot on the double ring

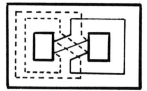

$$\begin{pmatrix} 2 & 3 & 1 \\ 0 & 2 & 0 \end{pmatrix}$$

$$\downarrow \text{II}$$

$$\begin{pmatrix} 2 & 0 & 2 \\ 1 & 0 & 2 \end{pmatrix}$$

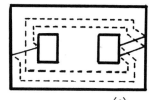

It therefore lies on the simple ring and has the symbol there $\begin{pmatrix} 2 \\ 3 \end{pmatrix}$. Less trivial examples may of course be given for p>2.

The final paper in this series is in effect a vindication
of Dehn's 1922 Breslau lecture, although it did not appear until
1938.   In it Dehn studies mapping classes by their action on
simple curve systems, for orientable surfaces of arbitrary genus   p,
finding that they are represented by linear transformations of
the space of   (6p-6)-dimensional integer vectors, and giving
a finite set of generators, which represent twist maps (now known
as Dehn twists).   The proof is by induction on the complexity
of the surface and the mapping, with a very complicated base step
which requires detailed analysis of spheres with up to 5 holes.
Perhaps because of its very demanding proof, the result went
unnoticed until it was rediscovered independently by Lickorish
[1962].   His proof is similar in concept to Dehn's but much
shorter, using twists to reduce complexity until a very simple
base step is reached.   Some further simplifications are made
in Birman [1977].   An impressive aspect of Lickorish's rediscovery is that he made it in conjunction with a rediscovery of
Dehn surgery, when Dehn himself had not connected these two ideas!

The work of Lickorish undoubtedly enables us to skip many
of the details of Dehn's proof, but it does not replace the whole
paper.   Dehn had another idea which also went unnoticed, even
when it was rediscovered in the unpublished, but famous, Thurston
[1976].   This was the idea of studying the mapping class group
by its action on the space of simple curve systems.   As mentioned
above, Dehn took this idea far enough to represent mapping classes
by linear transformations of a   (6p-6)-dimensional space of integer

vectors, but not far enough to understand the geometric meaning of this space.

Thurston [1976] shows how simple curve systems can be interpreted as "rational points" in the boundary of a space of surfaces of genus p, known as Teichmüller space T_p. $\overline{T}_p$ is defined as the space of hyperbolic structures on a surface of genus p, and thus requires a return to the hyperbolic geometry Dehn had abandoned in his quest for purely combinatorial proofs. Fricke and Klein [1897] found that $\overline{T}_p$ has dimension 6p-6, while Teichmüller [1939] found a natural metric on $\overline{T}_p$, under which the mapping class group acts on T_p as a group of isometries. The action extends continuously to Thurston's boundary points, which arise from degeneration of the hyperbolic structure, and the simplest type of degeneration yields simple curve systems. Thus Dehn's action of the mapping class group on simple curve systems can be viewed as the restriction of its action on the whole of $\overline{T}_p$ and its boundary. What Dehn did not know is that mapping classes are much easier to understand when they act on this large space, because $\overline{T}_p$ is in fact a topological ball.

Dehn was not able to give a finite set of defining relations for the mapping class group of arbitrary genus p. This was first done by McCool [1975], using algebraic methods, while a more geometric proof was given by Hatcher and Thurston [1980].

REFERENCES

J. Birman [1977] : The algebraic structure of surface mapping class groups. Discrete Groups and Automorphic Functions (Ed. W.J. Harvey), Academic Press, 163-198.

R. Fricke & F. Klein [1897] : Vorlesungen über die Theorie der Automorphen Funktionen I, Teubner, Leipzig.

A. Hatcher & W. Thurston [1980] : A presentation for the mapping class group of a closed orientable surface. Topology 19, 221-237.

W.B.R. Lickorish [1962] : A representation of orientable, combinatorial 3-manifolds. Ann. Math. 76, 531-540.

J. McCool [1975] : Some finitely presented subgroups of the automorphism group of a free group. J. Algebra 35, 205-213.

O. Teichmüller [1939] : Extremale quasikonforme Abbildungen und quadratische Differentiale. Abh. Preuss. Akad. Wiss., math.-naturw. Kl. 22, 1-197. Collected Papers, Springer-Verlag, 1982, 335-531.

W. Thurston [1976] : On the geometry and dynamics of diffeomorphisms of surfaces I (preprint).

THE GROUP OF MAPPING CLASSES

(The arithmetic field on surfaces)

CONTENTS

Introduction   259

§1.   Generalities on self-mappings of surfaces   261

  a) The group of mappings and the group of mapping classes

  b) Different types of mappings

  c) Homotopy with fixed and movable boundary

  d) Indicatrix

§2.   The self-mappings of the one-, two-, and three-holed spheres   264

  a) One-holed sphere

  b) Two-holed sphere

  c) Three-holed sphere with movable and fixed boundaries

§3.   Special mappings of the four-holed sphere   273

§4.   Self-mappings of the torus and the one-holed torus   276

  a) Torus

  b) One-holed torus

§5.   The arithmetic field on the two- and three-holed spheres

and the torus   278

a) General

b) Two-holed sphere

c) Torus.   Orientation of curves

d) Transformations of the arithmetic field on the torus by

mapping classes.   Generation and relations.

e) Three-holed sphere.   Arithmetic field.   Invariance under

homotopic transformations

§6.   The arithmetic field or the curve systems on the one-holed

torus   298

a) Introduction of the field

b) Transformation by mappings.   Examples.

c) One-holed torus with fixed boundary.   Connection with the

trefoil knot

§7.   Arithmetic field on the four-holed sphere   311

a) System of closed curves on the four-holed sphere

b) Mappings of the four-holed sphere and the action on the

arithmetic field

c) Derivation of invariants of a curve system from the arithmetic

presentation

d) Orientation.   Examples

e) Geometric presentation of the symbols.   View of higher cases

f) Curve systems on the four-holed sphere with endpoints on a

boundary

      1) Normal form

      2) Arithmetic field

  g) Four-holed sphere with fixed boundaries

      1) Derivation of a relation

      2) Application to the two-holed torus with fixed boundaries

      3) Twists along the boundary of singly-bounded surfaces

      4) Twists along separating curves on closed or singly-
         bounded surfaces

§8.   Five-holed sphere   335

  a) Coordinate systems

  b) Presentation of a system of closed curves

  c) Reduction of symbols. Generation of mappings

§9.   Generation of the mapping classes for the sphere with n holes 342

      1) - 5) Lemmas

      6) Generation with the help of complete induction

      7) The five-holed sphere as an example

      8) Direct exhibition of generators on the basis of
         a cyclic ordering of boundaries. The number of generators

§10.   Generation of the mapping classes for every orientable surface 353

      1) and 2) Lemmas

      3) Generation with the help of complete induction

      4) Double and triple torus as examples

      5) Direct exhibition of generators on the basis of a normal
         representation of the surfaces. The number of generators.

6) Arithmetic field in the general case.

Introduction

In combinatorial topology, topological concepts are represented by arithmetic concepts. Thus, in principle, all problems of combinatorial topology are reduced to arithmetic problems. However, this reduction is of no use for the resolution of the problems in most cases, because the corresponding arithmetic problems have little relation to known results or methods. This is especially true of all problems in which homotopic transformations are considered non-trivial or, as one can also say, in which the exceedingly numerous and hard to visualize constructions of simply connected polyhedra of different dimensions come into play. In an earlier work[*] I have attempted to represent this construction arithmetically in an understandable way for two-dimensional polyhedra. In doing so, I showed that homotopy problems yield arithmetic problems which fall outside the extensive domain of group theory. They concern more general operations which are difficult to investigate, the totality of which I have covered by the name "games".

The situation is quite different when one investigates problems in which homotopic transformations are neglected. Here one can greatly benefit from methods and results of group theory in many cases. For manifolds, even those of higher dimension, the

---

[*] Über kombinatorische Topologie, Acta math. **67** (1936), 123-168 (Translator's note).

fundamental group of POINCARÉ opens a way to attack such problems.
For two-dimensional manifolds, the theory of mappings in particular
has been worked out with great success in recent work which
collects all mappings which differ only by homotopies into a class
(see §1).

In the present work the mapping classes are represented as
operations in an arithmetic field on the surface. The individuum
of such a field cannot be a point on the surface, because a point
can be sent to any other point by a homotopy. In fact we choose
the individuals to be arithmetic representations of certain curve
systems which are partitioned into sets which collect together all
systems related by homotopies. Each such set is represented by
one curve system, which is uniquely determined by a number of
integers. One can consider this number to be the dimension of the
arithmetic field. The sequence of integers which determines a
particular curve system or the set of systems it represents is the
individuum of the arithmetic field. The mapping classes induce
transformations of this arithmetic field which are linear transformations related to modular substitutions.

This arithmetic representation makes it possible to solve a
series of simple mapping problems. The present work uses these
solutions to seize hold of more general problems, for which the
arithmetic field and its transformations are still only used in a very
rudimentary way. In this way it is possible to give a finite set

of generators for the mapping class group of an arbitrary surface. These generators are all similar types of mapping of the surface onto itself, namely twists along certain curves, which are simply related to usual representation of surfaces of higher genus.

This is a settlement of sorts, but the theory of transformations of the arithmetic field is very incomplete. A continuation in this direction would be of great significance. Certainly, with the presentation of the transformation formulae, the mapping class group is "represented" by linear transformations. But perhaps this representation will also make possible an attack on the conjugacy problem for mapping class groups, which has been the main theme of previous works in this field, from a new side.

Because of this possibility, the present work is to be considered only of a preparatory character. This character is also expressed in the detail devoted to simple things and in the fact that many known results are derived, either as a necessary basis for the method, or as useful examples.

§1.

Generalities on self-mappings of surfaces

a) The mapping Φ of a surface onto itself is given by a one-to-one correspondence between the vertices, edges and cells of one decomposition Σ_1 of the surface and another decomposition Σ_2

such that corresponding vertices lie on corresponding edges and the latter lie in the boundaries of corresponding cells. One can write $\Phi(\Sigma_1) = \Sigma_2$. However, Φ maps an arbitrary decomposition Σ of the surface onto a determinate decomposition Σ'. This is because Σ results from Σ_1 by internal subdivision. The internal subdivision is sent by Φ to a (topologically) determinate subdivision of Σ_2. We can therefore give the mapping for an arbitrary Σ and thus obtain $\Phi(\Sigma) = \Sigma'$. In particular, we obtain $\Phi(\Sigma_2) = \Sigma_3$ in this way, and hence $\Phi^2(\Sigma_1) = \Sigma_3$, and by continuing we obtain all powers of Φ. If Ψ is another mapping, then by the above we can also apply Ψ to Σ_2 and thus obtain $\Sigma_4 = \Psi(\Sigma_2) = \Psi\Phi(\Sigma_1)$. We can therefore compose the mappings of the surface, and the mappings of the surface form a group.

However, it would be impractical to consider this group directly, since it cannot be generated by a finite number of elements. One therefore considers a quotient group, which corresponds to a division of the mappings into classes : one reckons the mapping Φ to belong to the identity class if $\Phi(\Sigma_1) = \Sigma_2$ results from Σ_1 by a homotopy (that is, by deformation). Correspondingly, two mappings belong to the same class when one is the product of the other with a mapping in the identity class. The mappings in the identity class constitute a normal subgroup of the group of mappings. Because, if E is such a mapping, then $\Phi E \Phi^{-1}$ is the corresponding mapping on the polyhedron $\Phi(\Sigma)$ resulting from Σ, and hence also homotopic to the identity. The quotient by this normal subgroup, the mapping class group is the subject of our investigations.

b) A non-identity mapping can map some decomposition onto itself, in case the decomposition admits a non-identity mapping onto itself. Such a mapping is always of finite order, i.e. it has a power which not only belongs to the identity class, but is itself the identity mapping. Other mappings are not of this kind and have infinite order, so that none of their powers even belong to the identity class. Finally, there may also be a mapping of which a finite, say the nth, power belongs to the identity class, without any mappings in its class having an nth power equal to the identity. I.e. we would have $\phi^n = E$, where E is a deformation, while $(E'\phi)^n$ is different from the identity, no matter which deformation E' is chosen. We shall give examples of the first two kinds of mapping in the §§ which follow. Whether there are any of the third kind is doubtful.[*]

c) For bounded surfaces one can choose the identity class in different ways. One either lets the identity class consist of mappings which are homotopic to the identity without being the identity on the boundaries, or else one requires that each such mapping leave the boundaries pointwise fixed. This second, smaller, identity class is also a normal subgroup in the group of mappings. But when one uses it as a basis, one is restricted to mappings which leave the boundaries pointwise fixed. These mappings are subdivided more finely by their identity class than the group of general mappings

---

[*] Nielsen showed there are none in Acta math. **75** (1942), 23-115. (Translator's note.)

is subdivided by the larger identity class.

If "<u>punctured</u>" surfaces are under consideration, i.e. if each
boundary is contractible to a point, then the two identity classes
coincide.   Because each deformation leaves each boundary pointwise
fixed, since it consists of only a single point.   In addition to
the mappings which leave the boundaries fixed one may also mention
the special mappings which permute the boundaries (punctures).

d) A mapping sends an oriented curve to another curve with a
determinate orientation.   A mapping (not in the identity class)
either reverses or preserves the indicatrix.   The group of mapping
classes which preserve the indicatrix is a normal subgroup of the
group of general mapping classes.   When the contrary is not expressly
stated in what follows, mappings are assumed to be <u>indicatrix</u>
<u>preserving</u> (orientation preserving).

§2.

Self-mappings of the one-, two- and three-holed sphere

We shall call a sphere with n holes an n-holed sphere and
denote it by L_n.

a) It is practical to begin with direct consideration of a
few quite simple mapping problems, before considering the arithmetic
field on surfaces.   The mapping class group of the <u>one-holed sphere</u>

(the disc) is the identity group.   All mappings belong to the class
of the identity, whether the boundary is taken to be fixed or not.
The proof proceeds by successive homotopies from one subdivision to
another (topologically the same).

b) The mapping class group of the two-holed sphere (the cylinder)
with non-fixed boundaries is the group of order two (exchange of the
boundaries), but if boundaries are not to be exchanged (though their
points are not fixed) the group is the identity.   With pointwise
fixed boundaries the mapping class group is the infinite cyclic
group.   This happens because a point Y_1 of one boundary can be
connected to a point Y_2 of the other boundary in infinitely many
different ways (see the connections $Y_1 Y Y_2$ and $Y_1 Z Y_2$ in Fig. 1)

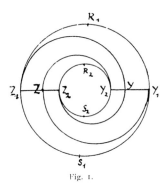

Fig. 1.

Fig. 1 represents a mapping which generates this cyclic group.   It
consists in the pointwise correspondence between the polygon

$Z_1 R_1 Y_1 Y Y_2 R_2 Z_2 Z$ and the polygon $Z_1 R_1 Y_1 Z Y_2 R_2 Z_2 Y$, and the polygon $Z_1 S_1 Y_1 Y Y_2 S_2 Z_2 Z$ and the polygon $Z_1 S_1 Y_1 Z Y_2 S_2 Z_2 Y$. This correspondence leaves the boundary points fixed (only the points Y and Z are exchanged). The polygons form two topologically equal decompositions of the cylinder. We shall call this mapping a _twist_. When the boundaries are fixed the twist is not a homotopy. The proof proceeds by first considering a simple arc $Y_1 Y Y_2$, connecting one boundary to the other, and its image $Y_1 'Y'Y_2'$. If the boundaries are allowed to move under homotopies, then all these connecting lines are homotopic to each other. But if the boundaries are pointwise fixed, in which case $Y_1' \equiv Y_1$, $Y_2' \equiv Y_2$, then there are infinitely many non-homotopic connecting lines, which are determined up to homotopy by the number of times they loop around one boundary in the positive or negative sense. This number is given by the algebraic sum of the positive or negative crossings of a connecting line $Z_1 Z Z_2$ which does not meet $Y_1 Y Y_2$. This number associated with $Y_1 Y' Y_2$ is invariant under homotopy, but the twist Δ shown in Fig. 1 raises or lowers it by one. This is easy to see when one deforms the mapping shown in Fig. 1, i.e. when one replaces it by a mapping Δ' in the same class. In fact (see Fig. 2) we take a point Y_1' on $Y_1 Y_2$ in the neighbourhood of Y, and a point

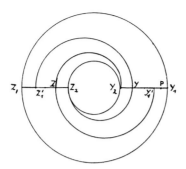

Fig. 2.

Z_1' on Z_1Z_2 in the neighbourhood of Z. We then replace the connecting line Y_1ZY_2 by $Y_1Y_1'ZY_2$ and the connecting line Z_1YZ_2 by $Z_1Z_1'YZ_2$. We can use a homotopy to arrange that the mapping remains the identity on the segments Y_1Y_1' and Z_1Z_1'. If we now have an arbitrary connection between Y_1 and Y_2, then we move all its points of intersection with Z_1Z_2 and Y_1Y_2 homotopically onto the segments Z_1Z_1' and Y_1Y_1' respectively; the connection between the last point of intersection on Z_1Z_1' and Y_2 can then be deformed into a connection between this point of intersection and a point P on Y_1Y_1' without any other intersections with Z_1Z_2 or Y_1Y_2, together with the polygonal path $PY_1'YY_2$. The mapping Δ' then deforms the segment $PY_1'YY_2$ into the curve $PY_1'ZY_2$ and we therefore get one more intersection point on Z_1Z_2, as claimed, because the remaining part of the deformed connecting line can be assumed fixed under the mapping.

A line connecting Y_1 and Y_2 which has the number zero is homotopic to $Y_1 Y Y_2$. By a suitable number of repetitions of the mapping Δ or its inverse, $Y_1 Y_2$ can therefore be converted into a line homotopic to an arbitrary line connecting Y_1 and Y_2. But the correspondence between $Y_1 Y Y_2$ and $Y_1' Y' Y_2'$ or $Y_1 Y' Y_2$ respectively completely determines the <u>class</u> of the mapping, because cutting along $Y_1 Y Y_2$ converts the two-holed sphere into the one-holed sphere, for which all mappings belong to the class of the identity. Thus the mapping class group is the identity in the first case (movable but not exchangeable boundaries) and the infinite cyclic group in the second (fixed boundaries), as asserted at the beginning, and it is generated by the mapping Δ. This group induces the group of translations in the numbers representing the lines connecting the boundaries.

I have presented the proof in detail because here we have the simplest manifestation of curves being given (up to homotopy) by numbers, and transformations of these numbers giving the group of mapping classes.

c) For the <u>three-holed sphere</u> (L_3) the group of mapping classes with movable boundaries is the <u>symmetric group of permutations of the three boundaries</u>[1], and the group with movable but not permutable

---

[1] In agreement with MAGNUS, Math. Ann. 109.

boundaries is the <u>identity</u>[1].   The proof proceeds by considering

two (simple) connections between the same two boundaries and

showing that they are homotopic under the stated assumptions.   We

use three connections AB, CD and EF to divide the sphere into

two discs (see Fig. 3).   We shall prove that each connection V

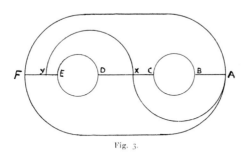

Fig. 3.

between A and B is homotopic to AB on L_3.   The connection

V is determined by its successive intersections with AB, CD and

EF.   We can assume that no consecutive intersections lie on the same

one of these segments, because the corresponding piece of V

could then be deformed into a piece on AB, CD or EF, and then

a further displacement of V into V' would   result in two

fewer intersection points.   In the same way we can arrange that

first intersection point does not lie on AB.   If it lies on

CD, say at X, then all other intersections lie on AB and XC,

because an intersection Y on EF would imply that all later

---

[1] This is the basis for the works on curve systems on surfaces by
DEHN, Autogr. Vortrag, Breslau 1922 and R. Baer, Journ. f. Math.,
vols. 156, 160.

intersections lay on YE and DX, and V then could never arrive at the point B. Hence in this case V makes a multiple circuit round the boundary BC, which can be reduced by deformation so that V no longer has intersections with the three connecting segments, and hence it bounds a disc on L_3 in conjunction with AB, i.e. it is homotopic to AB. The remaining case is where the first intersection X of V lies on EF (see Fig. 4). If the next intersection then lies on DC, the only possibility

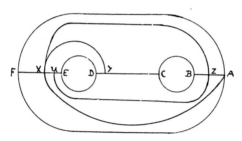

Fig. 4.

for V is to make circuits around the boundary CB, which can be ruled out. We can therefore assume that the next intersection Z lies on AB. Then the third intersection can again lie on DC, and we have circuits around the boundary CB, which we can rule out. Or else the next intersection U lies on XE. For the same reason we can now assume again that all later intersections lie alternately on AB and FE. We then have circuits around the boundary AF, which can likewise be ruled out. This completes the proof that V is homotopic to AB. But we also see from the proof that, with fixed boundaries, V is always homotopic to a number of circuits of the boundary AF followed by circuits

of the boundary BC. We note that it also follows that three

disjoint lines connecting pairs of the boundaries can be deformed

by a homotopy with movable boundaries into AB, CD and EF.

This follows immediately when one deforms V into AB and then

considers the L_2 resulting from L_3 by cutting along AB and

the boundary connections on it.

An arbitrary mapping Φ of L_3 with movable boundaries may

carry AB into V. A homotopy H may likewise carry AB into

V. Then the given mapping Φ differs from H by a mapping Ψ of

the two-holed sphere which results from L_3 by cutting along

AB, where the points A and B remain fixed. But each mapping

of a two-holed sphere, under which at least one boundary is movable,

is a homotopy. Thus Ψ is a homotopy of the two-holed sphere under

which the two segments corresponding to AB remain fixed.

Consequently, Ψ is also a homotopy of L_3. Thus Φ differs

from H only by a homotopy and it belongs to the class of the

identity. This completes the proof that the group of mapping

classes of L_3 with movable boundaries is the identity.

<u>The group of mapping classes of L_3 with fixed boundaries</u>

<u>is the free abelian group with three generators.</u> The three

generators are the twists along the three boundaries. A twist

along a boundary curve results from a twist of the two-holed sphere with fixed boundaries, on the surface itself the twist curve

lies between the boundary itself and a parallel curve.   In our

case we shall choose the three parallel curves to be disjoint.

The proof of our assertion follows easily from the foregoing.

Because the mapping Φ of L_3 may carry AB, the line connecting

the first and second boundaries, onto V.   As we have seen above,

we can convert AB into a curve $\bar{V}$ homotopic to V (with

fixed boundaries) by twists Δ_1 and Δ_2 along the first and

second boundaries.   I.e., when n_1 twists Δ_1 and n_2 twists

Δ_2 are needed the mapping[1] $\Psi \equiv \Delta_1^{-n_1} \Delta_2^{-n_2} H'\Phi$ carries AB into

itself, where H' is a homotopy which carries V to $\bar{V}$.   If we

again cut the L_3 along AB, the result is an L_2.   But a mapping

Ψ of L_3 which is the identity on AB and which by hypothesis

leaves the three boundaries of the L_3 fixed is a mapping which

carries L_2 onto itself with fixed boundaries, and hence by b)

it belongs to the class of a suitable power of the twist Δ_3 along

the third boundary.   Thus

$$\Delta_1^{-n_1} \Delta_2^{-n_2} H'\Phi = \Delta_3^{n_3} H'',$$

where H'' is a homotopy of L_2 with fixed boundaries, and hence

also a homotopy of L_3.   Hence, since the homotopies form a

normal subgroup of the group of mappings,

$$\Phi = H'^{-1} \Delta_2^{n_2} \Delta_1^{n_1} \Delta_3^{n_3} H'' = H \Delta_2^{n_2} \Delta_1^{n_1} \Delta_3^{n_3},$$

---

[1] The order of operations is always the right to left order of the symbols representing the operations.

i.e. the twists along the three boundaries generate the mapping class group.

The fact that these twists commute with each other follows from the assumption that the regions in which they occur are disjoint. There are no further relations between these three twists. This follows easily. If, say, $\Delta_1^{n_1} \Delta_2^{n_2} \Delta_3^{n_3} = 1$ in the mapping class group, then we consider a connection v_{12} between the first and second boundaries. With a suitable choice of the twist strip along the third boundary, v_{12} is not changed at all by Δ_3, and then $\Delta_1^{n_1} \Delta_2^{n_2}$ only carries it into a homotopic curve if n_1 and n_2 are both zero. But since a mapping belongs to the identity class only when each connecting line v_{12} is carried to a homotopic line, $\Delta_1^{n_1} \Delta_2^{n_2}$ can only belong to the identity class in case $n_1 = n_2 = 0$. We obtain the condition $n_1 = n_3 = 0$ similarly. Thus $\Delta_1, \Delta_2, \Delta_3$ generate a free abelian group.

3.

Special mappings of the four-holed sphere

The four-holed sphere L_4 possesses mappings with movable but not permutable boundaries which do not belong to the identity class. We denote the four boundaries by r_1, r_2, r_3, r_4 (see Fig. 5)

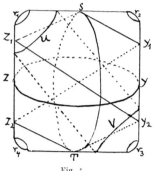

Fig. 5.

and consider three closed curves YZ and ST and UV which
respectively separate r_1 and r_2 from r_3 and r_4, r_1 and r_4
from r_2 and r_3, and r_2 and r_4 from r_1 and r_3. We now
consider (by §2) the twists along ZY, ST and UV. These are
mappings which do not belong to the class of the identity, even with
movable boundaries, because the twist along ZY, e.g., converts ST
to the curve $f \equiv SY_1Z_2TY_2Z_1$ (see Fig. 5) which is not homotopic
to ST. One sees this most easily by considering the <u>fundamental</u>
<u>group</u> of L_4. The latter is the free group on the three generators
r_1, r_3, r_4 and within it ST corresponds to a conjugate of r_1r_4,
while f corresponds to a conjugate of $r_1r_3r_4r_3^{-1}$. But the latter
element is not a conjugate of r_1r_4 in the free group. Thus f
is not homotopic to ST and the (twist) mapping which carries ST
to f is not a homotopy. One reads the representation of f in
the fundamental group from the course of the two curves in the
polygon formed by cutting L_4 along connections to the four

boundaries from a point O (see Fig. 6).

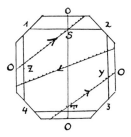

Fig. 6.

The same holds for the twists along ST and UV. We shall see later (§7), that one <u>can generate a mapping of</u> L_4 <u>with movable but not permutable boundaries in each class by means of the twists along</u> XY <u>and</u> ST, <u>so that the mapping class group in question is generated by these two mappings.</u> It is obvious that the twist along ST results from that along XY by conjugation with a suitable permutation of the boundaries.

In the case where the boundaries are immovable, the mapping class group results from the above by adding the twists along the four boundaries. These commute with each other and with any mappings which leave the boundaries fixed, in particular with the twists along ST, YZ and UV, and there is a simple relation between the four boundary twists and the three twists along XY, ST and UV (see §7g).

§4.

Mappings of the torus and the one-holed torus

a) The mapping class group of the torus R has long been known from function theory.   However, we shall use our method to treat it in connection with other problems.   We first give two special mappings of R, namely the twists along two closed curves which cross at one point, say a and b (see Fig. 7).

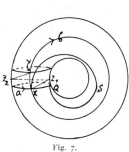

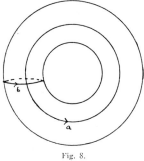

Fig. 7.   Fig. 8.

We denote the twists by Δ_a and Δ_b. Δ_a replaces the connection XY by XZ_1Z_2Y, and Δ_b replaces the connection XQ by XSQ.   Then if Σ is the   decomposition of R shown in Fig. 7, generated by a and b, one easily sees by successive application of the transformations in question that $\Delta_a\Delta_b^{-1}\Delta_a(\Sigma)$ is given up to homotopy by Fig. 8.   But this figure represents a rotation of the topologically regular torus polyhedron of Fig. 7, with edges a and b and a vertex χ, about its vertex, and hence $\Delta_a\Delta_b^{-1}\Delta_a$ belongs to a mapping class of finite, in fact fourth, order. We obtain $(\Delta_a \Delta_b^{-1} \Delta_a)^4 = 1$. Of course $(\Delta_b \Delta_a^{-1} \Delta_b)^4 = 1$ similarly. We shall see later that representatives of all mapping classes of R can be generated by Δ_a and Δ_b.

Δ_a and Δ_b are also mappings of the bounded torus R_1, because the twists affect only the neighbourhoods of a and b.

The mapping class group for the closed torus R is endowed with an orientation by the mapping of a and b. Namely, if ϕ and ϕ' both send the (oriented) a and b to a' and b', then $\phi'^{-1}\phi$ carries the curves a and b onto themselves with preservation of orientation. We have to prove that $\phi'^{-1}\phi$ belongs to the class of the identity. First we deform $\phi'^{-1}\phi$ so that it becomes the identity on a and b. This is possible because $\phi'^{-1}\phi$ maps the curves a and b onto themselves with preservation of orientation. If we now cut R along a and b, the result is a disc which $\phi'^{-1}\phi$ maps onto itself with fixed boundary. But such a mapping is homotopic to the identity with fixed boundary, and a homotopy of the disc with fixed boundary is also a homotopy of R. Thus $\phi'^{-1}\phi$ is homotopic to the identity mapping and therefore belongs to the identity class. One observes that a homotopy of the disc with non-fixed boundary in general does not correspond to a transformation of R at all, when the four points on the boundary of the disc which correspond to the intersection of a and b are not sent to themselves.

b) It follows similarly that the mapping class group of the one-holed torus R_1 with movable boundary is the same as that for the closed torus.   Because, if we make exactly the same operations as in the above section, then $\phi'^{-1}\phi$ maps a two-holed sphere with one fixed and one movable boundary onto itself.   But the mapping class group for this figure is also the identity. Because, as we saw in §2, each mapping of the two-holed sphere is deformable with fixed boundary so that the neighbourhood of one boundary is unmoved while the neighbourhood of the other is twisted. If the second boundary is now taken to be movable, then the mapping may be deformed into the identity.   Consequently $\phi'^{-1}\phi$ belongs to the identity class.

§5.

The arithmetic field on the two- and three-holed sphere and on the torus

a) The arithmetic fields on the surfaces result from systems of curves on the surfaces.

Our general principle of representation is to represent homotopic curves identically.   Also, we shall take the points of the arithmetic field to be, not single open or closed curves, but whole systems of such curves without double points or intersections. Moreover, we shall omit from such systems all curves which are contractible to a point or to a proper part of the boundary. It

is necessary to observe these conditions in order to make the

representation at all simple. We recall for a moment the usual

representation of surface curves by elements of the fundamental

group in order to see where it is different : the elements of the

fundamental group correspond to single closed curves with arbitrary

singularities. But the curves are represented uniquely only when

a fixed point on the surface is given as their initial point, and

when they are assigned orientations ; curves for which these extra

specifications are not made correspond to all the elements which result

from a single element by conjugation. In our representation, the

curves or curve systems are represented by a fixed number of integers,

corresponding to the dimension of the arithmetic field. On the

other hand, the element of the fundamental group representing a

curve is given by an arbitrarily long sequence of integers, the

exponents of the different generators. Our simpler representation

makes the action of a self-mapping of the surface on these number

complexes much easier to follow, and brings it into line with known

arithmetic transformations.

b) Curve systems on the two-holed sphere

When the boundaries are movable, a system of lines connecting

the boundaries is determined by their number, say n, alone.

Because any two connecting lines are homotopic when the boundaries

are movable.   Any two systems of n disjoint connecting lines
are always homotopic.   Also, one can pair one line of one system
with an arbitrary line in the other system.   However, these
connecting lines do not exhaust the possibilities for non-contractible
curves.   There are still the closed curves parallel to the
boundaries.   These become represented when we now <u>fix the boundaries</u>
<u>under the homotopy</u> : on each boundary we choose n fixed points,
and we also choose a <u>normal line</u> v connecting the boundaries and
a <u>positive side</u> of v.   Then we consider the algebraic sum δ of
the positive and negative crossings of v as we traverse the n
connections from the "first" boundary to the "second".   By means
of a homotopy with fixed boundaries we can arrange that all
connections, since they are disjoint by hypothesis, cross v in the
same sense.   If, first, |δ| is smaller than n, then there are
exactly |δ| connections which cross v.   They result from
connections which do not cross v by means of a single twist of a
boundary through one turn (see Fig. 9).   Hence by §2b there

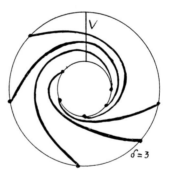

Fig. 9.

can be no homotopy with fixed boundary which changes this number

δ. However, if $\delta = \nu + \mu n$ with $|\nu| < n$, where ν has the same

sign as μ, then the system of connections results from a single

twist of ν connections followed by a μ-tuple twist of all n

connections. Since, by §2b, the number of windings of each

individual connection is not changed by homotopy with fixed boundary,

the number δ is independent of such homotopies.

The number pair $\binom{n}{\delta}$ determines a system of connections up

to homotopy with fixed boundary. We now associate the symbol $\binom{0}{\delta}$

with a system of $|\delta|$ oriented curves parallel to the boundaries,

with the sign of δ describing whether they cross the normal connection

ν in the positive or negative sense ; then the correspondence

between curve systems and symbols $\binom{n}{\delta}$ is complete, and unique up

to homotopy with fixed boundaries.

The mappings of the two-holed sphere onto itself which do not

exchange boundaries are the twists. They transform the symbol

$\binom{n}{\delta}$ (by §2b) into $\binom{n}{\delta + kn}$, where k is a positive or negative

integer. The collection of transformations of these symbols, when

$n \neq 0$, makes up a group isomorphic to the mapping class group of

the two-holed sphere. When $n = 0$ the twist transforms the symbol

into itself, so the group of transformations of this symbol is the

trivial group. In this representation only the closed curves have an

orientation independent of the symbols.

c) Curve systems on the torus

We divide the torus into two two-holed spheres L_2 and L_2'
by means of two disjoint non-separating curves a_1 and a_2 (Fig. 10).

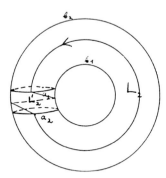

Fig. 10.

An arbitrary curve system on the torus, apart from curves parallel
to a_1, consists of connections between n points on a_1 with n
points on a_2, in both L_2 and L_2', with any self-connections of
a_1 or a_2 in L_2 or L_2' being removed by homotopies. By homotopies with movable boundary, say in L_2, these n points can be
carried to n given points on each of the boundaries, and likewise
the connections in L_2 to given connections. Even then, the
association of the n fixed points and the n fixed connections
to the n given connections is determined only up to cyclic interchange. This cyclic interchange corresponds to a homotopy of the torus.

If the connections in L_2 are now fixed then L_2' is transformable only by homotopies with fixed boundaries. The n connections of the n fixed points on the two boundaries in L_2' are therefore, by b), represented by a number pair $\binom{n}{\delta}$, where δ is the algebraic sum of the number of crossings, in the positive or negative sense, over a normal connecting a_1 and a_2, if we always take the curve system inside L_2' to run from a_1 to a_2. We can take such a normal to be, say, the piece in L_2' of a curve b_1 which cuts a_1 once (see Fig. 10) and, when one side of b_1 is chosen to be positive, a crossing of the normal from the negative to the positive side will be reckoned positive. By means of a homotopy, it can be arranged that all the crossings have the same sign, so that $|\delta|$ represents the number of crossings. We also take the piece of b_2 in L_2 parallel to the fixed connections in L_2, so that the curves of the system have exactly $|\delta|$ points of intersection with b_1. By means of homotopies we can also arrange that the crossings of curves in the system with a curve b_2 parallel to b_1 all have the same sign. Then the curve system can also be regarded as a system of connections of $|\delta|$ points on b_1 with $|\delta|$ points on b_2 in the two two-holed spheres L_2'' and L_2''' which result when the torus is cut along b_1 and b_2. Finally, we shall associate the symbol $\binom{0}{\delta}$ with a system of $|\delta|$ curves parallel to a_1 and a_2. We then have a number pair $\binom{n}{\delta}$ for each curve system on the torus, with $n \geqslant 0$ and δ arbitrary, positive, negative or zero. With the exception of curves parallel to a_1, we have given the curves no orientation, and this is also expressed by our symbol (n is only a number).   This has the consequence that a mapping of the torus onto itself is first determined by the images of three curves, since e.g. a_1 and b_1 can be mapped onto themselves with reversed orientation.   On surfaces of higher genus the problem has not previously been solved, since the orientation has not been represented arithmetically.   However, on the torus, the introduction of an orientation for curves can be made without difficulty : on the torus, all curves of a system are homotopic ; they are parallel curves, as follows immediately from topological considerations.   Namely, cutting along a curve c yields an L_2, and all simple, closed, non-contractible points on the L_2 are homotopic to the boundary curve c.   We give the curves of the system an orientation so that homotopic transformation of one curve in the system to another preserves orientation.   If one comes from a_1 to a_2 by traversing a piece of the curve system in L_2', then the same is true for any piece of the system in L_2', since we have assumed that each self-connection of a_1 or a_2 in L_2 or L_2' has been removed by a homotopy.   It is the same for the pieces of all other curves in the system.   In this case we give n   the positive sign.   If, on the other hand, one comes from a_2 to a_1 on traversing a curve piece in L_2', then n shall have the negative sign.   In addition — and anomalously with the above determination — δ shall have the positive sign when

the curve piece in a particular one of the pieces into which the
torus is divided by b_1 and b_2 goes from b_1 to b_2. If, in
Fig. 10, we take this part as the back side of the surface, then
the curve represented in the figure has the symbol $\binom{1}{1}$. If we
reverse the orientation, we obtain the symbol $\binom{-1}{-1}$ for this
curve. In general, reversal of orientation changes the symbol
$\binom{n}{\delta}$ to $\binom{-n}{-\delta}$. Finally, we shall associate the symbol $\binom{0}{\delta}$ with
a system of $|\delta|$ curves parallel to a_1, with the sign of δ
corresponding to the orientation.

d) Now we consider the effect of the <u>mappings</u> Δ_a and Δ_b
(§4), the twists along a_1 and b_1 in L_2' and L_2''' respectively,
on the symbol $\binom{n}{\delta}$. As we have said in §5c) and §2 about these
mappings and the mappings of a two-holed sphere, Δ_a changes the
number δ in $\delta + n$, and Δ_b changes n into $n + \delta$. Thus we can
use Δ_a and Δ_b, assuming $n > 0$, to effect the operation of the
<u>euclidean algorithm</u> on the symbol, and hence by a suitable sequence
of applications of $\Delta_a^{\pm 1}$ and $\Delta_b^{\pm 1}$ reduce $\binom{n}{\delta}$ to the symbol
$\binom{(n,\delta)}{0}^1$, which represents (n,δ) curves parallel to b_1 with
positive orientation, by the above prescription. As we have seen
in §4, we can also use Δ_a and Δ_b to reduce the system to (n,δ)
curves parallel to b_1 with negative orientation, or (n,δ) curves
parallel to a_1 with positive or negative orientation. This

---

[1]Where (n,δ) denotes the greatest common divisor of n and δ.

corresponds to the arithmetic fact that we can reduce $\binom{n}{\delta}$ to $\binom{\pm(n,\delta)}{0}$ as well as $\binom{0}{\pm(n,\delta)}$ by our arithmetic operations.

If the original curve system consists of a single curve, then $(n,\delta) = 1$. If we have two systems which each consist of a single curve, then we can reduce their symbols, by Δ_a and Δ_b or the euclidean algorithm, simultaneously to $\binom{1}{0}$ and $\binom{n_2}{\delta_2}$. $\binom{1}{0}$ is a curve parallel to b_1, and hence it cuts the second curve in $|\delta_2|$ points. If the two given curves have only one point of intersection, then $\delta_2 = \pm 1$, and by further application of Δ_b we transform the symbols for the two systems into $\binom{1}{0}$ and $\binom{0}{+1}$.

Now each mapping sends a certain oriented curve pair to the one represented by $\binom{1}{0}\binom{0}{1}$, and it sends the same pair, in which the second curve has the opposite orientation, to the pair represented by $\binom{1}{0}\binom{0}{-1}$. By §4a), each mapping which simultaneously sends $\binom{1}{0}$, $\binom{0}{1}$ and $\binom{1}{0}$, $\binom{0}{-1}$ to themselves belongs to the class of the identity. Thus two mappings which send the same curves to $\binom{1}{0}$, $\binom{0}{1}$ and $\binom{1}{0}$, $\binom{0}{-1}$ respectively belong to the same class. The transformations of the arithmetic field uniquely determine the mapping classes. On the other hand, when there is a mapping which carries a curve pair to $\binom{1}{0}$, $\binom{0}{1}$, then none of the mappings considered here will carry this curve pair to $\binom{1}{0}$, $\binom{0}{-1}$, otherwise one of the mappings would reverse orientation. We can therefore transform each curve pair which is sent to $\binom{1}{0}$, $\binom{0}{1}$ by a mapping, into $\binom{1}{0}$, $\binom{0}{1}$ by Δ_a and Δ_b also. Thus each mapping class contains an element generated by Δ_a and Δ_b : Δ_a and Δ_b generate the mapping class group of the torus.

In §4 we have given two relations for Δ_a and Δ_b. The following consideration gives us one more : the mapping $\Delta_a^{-1}\Delta_b\Delta_a^{-1}$ changes the curve a_1, with a certain orientation, into b_1 with a certain orientation, hence the twist Δ_a along a_1 is transformed by the operation $\Delta_a^{-1}\Delta_b\Delta_a^{-1}$ into a twist along b_1. Accordingly,

$$\Delta_a\Delta_b^{-1}\Delta_a \cdot \Delta_a^{-1} \cdot \Delta_a^{-1}\Delta_b\Delta_a^{-1} = \Delta_b$$

is a relation, easily checked by the representation of Δ_a and Δ_b as linear transformations. If we set

$$\Delta_a = \Sigma^{-1}T^{-1}, \ \Delta_b = \Sigma^{-2}T^{-1}$$

with the solution:

$$\Sigma = \Delta_a\Delta_b^{-1} \ , \ T = \Delta_a^{-1}\Delta_b\Delta_a^{-1}$$

in these relations then we obtain

$$\Sigma^3 T^2 = 1.$$

On the other hand, it follows from the relation

$$(\Delta_a^{-1}\Delta_b\Delta_a^{-1})^4 = 1$$

obtained in (§4) that

$$T^4 = 1.$$

Thus we have : the mapping class group is generated by two operations Σ and T which satisfy the relations

$$T^4 = \Sigma^3 T^2 = 1.$$

As is well-known, the group with these two relations between its generators is the homogeneous modular group, and hence the latter is the mapping class group of the torus. In fact, one easily obtains a normal form for the elements on the basis of the two relations, and then one proves, with the help of uniqueness of the continued fraction expansion, that the two relations are sufficient for the definition of the mapping class group, and hence this group is identical with the homogeneous modular group.

e) Curve systems on the three-holed sphere

We first consider only the unclosed curves. These connect either two different boundaries or else two different points on the same boundary. In the latter case we call it a self-connection of the boundary in question. Since we consider no curves which are contractible on a segment of the boundary, a curve connecting two points on the same boundary must separate the other two boundaries from each other, and the latter must therefore have no connection with each other. If n_1, n_2, n_3 are the numbers of endpoints of curves of the system on the three boundaries, then $n_1 + n_2 + n_3$ is even. Further, if $v_{ik} = v_{ki}$ is the number of connections between the ith and k^{th} boundary, then

$$n_i = \nu_{ik} + \nu_{i\ell} \quad \text{and} \quad \nu_{ik} = \frac{n_i + n_k - n_\ell}{2}$$

if there are no self-connections, or, what is the same thing, if $n_i + n_k$ is always greater than n_ℓ. If, however, say

$$n_i > n_k + n_\ell ,$$

then

$$\frac{n_i - n_k - n_\ell}{2} = \nu_{ii}$$

is the number of self-connections of the ith boundary. Moreover,

$$\nu_{ki} = n_k , \quad \nu_{\ell i} = n_\ell$$

in this case. Thus in each case the number of connections from boundary to boundary is <u>uniquely determined</u> by n_1, n_2, n_3 and <u>the whole system is determined up to homotopy, provided the boundaries are movable.</u> I.e. for given values n_i we can convert each system into one we call a normal system, by homotopies with movable boundaries. This is what we proved in §2c) for connections between different boundaries. However, it also follows easily for self-connections. Suppose that v_{11} is such a self-connection, say of the first boundary. Then there are disjoint connections v_{12} and v_{13} of the first boundary with the second and third which do not cut v_{11}, and a connection v_{23} of the second boundary with the third which cuts v_{11} once. We similarly associate with a second self-connection v_1' three connections v_{12}', v_{13}' and v_{23}' with the same properties.

However, when we carry v_{12}, v_{13} and v_{23} simultaneously to v'_{12}, v'_{13} and v'_{23} by a homotopy as in §2c), then v_{11} goes to a $\bar{v}_{11}$ which does not cut v'_{12} and v'_{13} and which cuts v'_{23} once. Thus we can convert $\bar{v}_{11}$, and hence also v_{11}, into v'_1 by a homotopy. However, if v_{11} is fixed, then v_{12} and v_{13} are convertible to v'_{12} and v'_{13} by homotopies which move the second and third boundaries.

In contrast to the situation on the two-holed sphere or the closed torus, the association of the given connections with the fixed connections can never be cyclically altered. Because in a series of disjoint connections of, say, the first boundary with the second (see Fig. 11) there are two distinguished connections

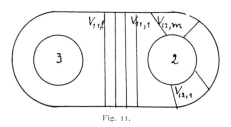

Fig. 11.

$v_{12,1}$ and $v_{12,m}$ which cut the L_3 into two parts — an L_1 on which the other connections lie, and an L_2. No other pair of connections has this property of excluding all other connections from the resulting L_2. Also, $v_{12,1}$ cannot be exchanged with $v_{12,m}$, because this would reverse the orientation of the surface. Similarly, there are two distinguished, non-exchangeable connections

$v_{11,1}$ and $v_{11,\ell}$ among the self-connections of a boundary (see Fig. 11).

All curve systems which contain no closed curves are represented by the number triple n_1, n_2, n_3 relative to the three boundaries. The systems with closed curves are obtained when we again fix the boundaries. To investigate this we use the same methods as for the two-holed sphere: we connect the three boundaries r_1, r_2, r_3 to each other by disjoint normal connections v_{12}, v_{23}, v_{31} and cut the v_{ik} by curves r_i' and r_k' parallel to r_i and r_k in six points altogether (see Figs. 12, 13). The normal connections and the boundary curves form two hexagons S_1 and S_2, and similarly the pieces of the normal connections and the parallel curves form hexagons S_1' and S_2' which are parts of S_1 and S_2 respectively. We now determine n_i fixed points on r_i and r_i', and in fact, if

$$n_k + n_\ell > n_i$$

(in Fig. 12, i = 1, k = 2, ℓ = 3) we place all points on r_i and r_i' in the hexagon S_1. However, if

$$n_k + n_\ell < n_i$$

(Fig. 13) then we put

$$\frac{n_i - n_k - n_\ell}{2}$$

points on r_i and r_i' in S_2, and

$$\frac{n_i + n_k + n_\ell}{2}$$

points in S_1. Now in the first case we connect the n_i points on r_i' with

$$\frac{n_i + n_k - n_\ell}{2}$$

points on r_k' and

$$\frac{n_i + n_\ell - n_k}{2}$$

points on r_ℓ' by the disjoint connections in S_1',

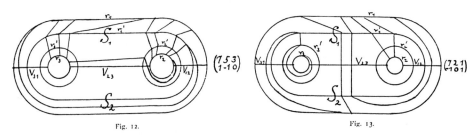

Fig. 12. Fig. 13.

in the second case we connect

$$\frac{n_i - n_k - n_\ell}{2}$$

suitable points on r_i', which lie in S_1', with the

$$\frac{n_i - n_k - n_\ell}{2}$$

points on r_i' which lie in S_2' by the disjoint lines which do not cut v_{ik} and $v_{\ell i}$ and which cut $v_{k\ell}$ only once, and the other n_k points on r_i' with the n_k points on r_k', n_ℓ points on r_i' with the n_ℓ points on r_ℓ', by the lines in S_1' which do not cut themselves and the v_{ik}. In such a system of connections, each system of connections between two boundaries, to the extent that it lies in S_1' and S_2', can be homotoped with <u>fixed</u> boundaries r_i but movable r_i'. Only the connections in the three $L_{2,i}$ between the r_i and r_i' now remain. We treat these exactly as above and determine δ_i as the algebraic sum of the number of crossings of v_{ik} in the positive and negative sense, where the positive side of v_{ik} is taken to be the one in S_1, say. The prolongation of connections from r_i to r_i' is fixed. This determines a symbol $\begin{pmatrix} n_1 & n_2 & n_3 \\ \delta_1 & \delta_2 & \delta_3 \end{pmatrix}$. <u>We shall show that it is unchanged by homotopies with fixed boundary</u>.

If <u>no self-connections</u> occur, then we can apply exactly the same considerations as we did above with the two-holed sphere : if, say, $\delta_1 = \nu_1 + \mu_1 n_1$, this means that the system of connecting lines of the ("undisturbed") normal system, which consists purely of v_{12} disjoint connections, is made up of ν_1 connections which circle r_1 <u>once</u> before being accompanied by the rest of the system in μ_1 circuits of r_1 in the same sense. This property, and with it δ_1, cannot be altered by homotopies, as we know from the investigation of connections between <u>different</u> boundaries on the three-holed sphere (§2c)). However, we must still consider the case of self-connections,

which was not investigated in §2c). We first remark that, as we saw
above, all self-connections of a boundary can be carried into
each other by homotopies with movable but not exchangeable boundaries,
however, the <u>endpoints cannot be associated arbitrarily</u>. Because
(see Fig. 14) the self-connection AB, which runs from A to B,
determines two different orientations for the two L_2's into which
it cuts the L_3. Since both L_2's contain non-exchangeable
boundaries, the two L_2's cannot be exchanged. Thus if A'B' is
another self-connection of the same boundary, then, assuming A'B'
gives the two L_2's the same orientations as AB, A'B' can be
carried to AB only by carrying A' to A and B' to B. When
A' goes to B and B' goes to A, then the orientations given by
AB and A'B' must be different.

Moreover, the connections between, say, the first and second
boundaries with the same endpoints can be converted into each other
by mappings with fixed boundaries. The self-connections of the
first boundary, on the other hand, divide into two classes (see Fig. 15).

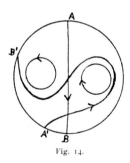

Fig. 14.

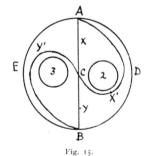

Fig. 15.

The connection $AXCYB \equiv v$ cannot be converted into the connection $AX'CY'B \equiv v'$ with the first boundary fixed, because the parts $ADB \equiv r_{11}$ and $AEB = r_{12}$ of the first boundary remain fixed. The closed curve made up of v and r_{11} is however homologous to r_2 on the L_3, whereas the closed curve made up of v' and r_{11} is homologous to r_3, hence no mapping with fixed boundary carries v into v'. If we again argue as in §2b), then we realise that all connections between A and B (other than those homotopic to a piece of the boundary, which are excluded), result from v or v' by circling the boundary r_1.

In the case that $n_1 > n_2 + n_3$, suppose first that $|\delta_1| < n_1$. For the sake of convenience we shall replace the normal connection v_{31} by v_{31}', where all fixed points on r_1, and the corresponding points on r_1' lie in one of the hexagons S_1 or S_2, say S_1, made up of v_{12}, v_{23} and v_{31}' as well as the r_i (see Fig. 16). The number δ_1, since it is defined by crossings of v_{12}, is not altered by this. Now there are again exactly δ_1 connecting segments in the $L_{2,1}$ between r_1 and r_1' which circle around r_1 once relative to v_{12} and v_{31}. As long as these segments belong to connections with other boundaries, the circling, as we already know, is not altered by homotopy. However, when one of the two segments belonging to a self-connection circles r_1, this self-connection cannot be converted into a self-connection without circling segments by mapping with fixed boundaries. Because (see Fig. 16) the undisturbed self-connection GM, together with the part of r_1 belonging to S_1 is homologous to r_3, but the disturbed self-connection G'M', together with the part of r_1 in S_1 is homologous to r_2. However, when both segments in $L_{2,1}$ belonging to a self-connection are replaced by circling connections, then the disturbed self-connection results from an undisturbed self-connection by a twist in $L_{2,1}$ (see Fig. 17). Here the word "undisturbed" means that the connection in question has the same

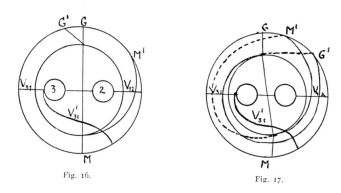

Fig. 16.   Fig. 17.

endpoints on r_1 and r_1', but it does not cut the segments v_{12} (and v_{31}') in $L_{2,1}$ (in Figure 17 the segments in $L_{2,1}$ which convert M'G' into an undisturbed connection are shown dotted). Altogether, we see that the number δ_1 of crossings of v_{12} is also not changed in this case by a homotopy with fixed boundary, because it represents a distinction between the given connections of the first boundary with other boundaries or itself and the undisturbed connections which cannot be changed by homotopy. And in fact there is a different distinction for positive and negative δ.

Finally, if $|\delta_1| > n_1$, say $\delta_1 = \nu_1 + \mu_1 n_1$, where μ_1 has the same sign as δ_1, then we have to add μ_1 twists of the whole system of connections in $L_{2,1}$, with the same sign as δ_1, to the alteration given by ν_1. Homotopy with fixed boundary does not change δ_1. Thus each system of connections uniquely yields the symbol $\begin{pmatrix} n_1 & n_2 & n_3 \\ \delta_1 & \delta_2 & \delta_3 \end{pmatrix}$ independently of homotopy with fixed boundaries.

When $n_i = 0$ (so that there is no connection with the ith boundary) we associate the number δ_i with $|\delta_i|$ curves parallel to r_i, with orientation corresponding to the sign of δ_i. Then each such symbol is also associated uniquely, up to homotopy, with a curve system connecting the n fixed points on the three boundaries, and conversely, each system of closed or non-closed curves on the three-holed sphere is represented by such a symbol. Incidentally, each such curve system divides into at most three classes of curves which are mutually homotopic with movable boundaries. The numbers of members of these classes are, for the different cases,

$$\frac{n_i + n_k - n_\ell}{2} \quad \text{or} \quad \frac{n_i - n_k - n_\ell}{2} \quad \text{and} \quad n_k \quad \text{or finally} \quad \delta_i \quad (\text{for } n_i = 0).$$

The mappings of L_3 onto itself are known to us; they result from twists along the three boundaries. They transform our symbol into

$$\begin{pmatrix} n_1 & n_2 & n_3 \\ \delta_1 + k_1 n_1 & \delta_2 + k_2 n_2 & \delta_3 + k_3 n_3 \end{pmatrix}$$

where the k_i are integers, positive, negative or zero.

In conclusion, we remark that <u>here</u> we can give the n_i <u>no signs</u>, unlike the case of the torus.

§6.

<u>The arithmetic field or the curve systems on the one-holed torus</u>

a) We divide the one-holed torus R_1^1 as we did the closed torus R_1 (see §5c)), by two disjoint closed non-separating curves a_1 and a_2, into a two-holed sphere $L_{2,a}$ and a three-holed sphere $L_{3,a}$ (see Fig. 18). A curve system in R_1^1 has say n points of

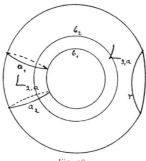

Fig. 18.

intersection with a_1, as well as a_2, and $2q$ endpoints on the boundary r. By means of homotopies with movable boundaries on

$L_{3,a}$, we can convert the curve system in $L_{3,a}$ into one with n fixed points on each of a_1 and a_2 and 2q fixed points on the boundary r, and fixed connections. The latter may be curves which do not cut the curves b_1 and b_2 or, in the case of self-connections of r in $L_{3,a}$, they may have one point in common with each of b_1 and b_2. In fact, this association of the given connections with the fixed ones is unique, as we saw for the three-holed sphere in §5e). Self-connections of a_1 or a_2 in $L_{3,a}$ cannot appear, since n < n+2q. However, there certainly can be self-connections of the boundary r, if q > n.

There still remain the connections in $L_{2,a}$ between the n points on a_1 and the n points on a_2. These are determined by a positive or negative number δ (see §5c)). As with the torus, we now give curves on R_1 which cut a_1 and a_2 equal orientations, and thereby give n a sign. However, not all curves are homotopic here. We must therefore arrive at agreement of sign somewhat differently so that all curves which cut a_1 cross it in the same sense. We can arrange this, because a curve which crosses a_1 more than once (apart from points removable by homotopies) always crosses it in the same sense, by circling around the curve b_1. (It is here that the determination of sign for higher genus fails.) This determination gives n a sign, positive for positive crossing and negative for negative crossing. At the same time, the determination gives all crossings of b_1 the same sense. For those of the latter which do not cut a_1, namely the self-connections of the boundary r in $L_{3,a}$, we determine the orientation so that the crossing of b_1 is in the same sense for all curves. In this way, δ is definable analogously as for the closed torus. Thus $\binom{n}{\delta}q$, where n and δ are arbitrary integers, and q is a natural number, gives us an arbitrary curve system on R_1, when we disregard curves parallel to the boundary. The only exception is the curve system represented by the symbol $\binom{n}{0}q$. In this case the orientation of the $q - |n|$ self-connections of the boundary is arbitrary.

In order to obtain the curves parallel to r as well, we must consider homotopies with fixed boundary (just as with L_3, §5e)) and introduce a rotation number for the boundary. This proceeds exactly as above, however we do not need it later and we shall therefore leave these curves out of consideration. Later in this paragraph, though, we shall give a theorem on the mappings with fixed boundary.

The arithmetic field we have introduced for R_1 is three-dimensional and its points are all oriented curve systems on R_1 minus curves parallel to r.

b) <u>Transformation of the arithmetic field or the symbols</u> $\binom{n}{\delta}q$ <u>by mappings of the one-holed torus onto itself.</u>

The mappings are the same as for the torus (see §4b)) ; the

mapping classes are generated by the twists, Δ_a along a curve a

parallel to a_1, and Δ_b along a curve b which cuts a once.

These mappings induce the operations of the euclidean algorithm in

the symbols $\binom{n}{\delta}$ of the torus R. How do these mappings act on

the symbols $\binom{n}{\delta} q)$? The action of Δ_a is the same as on the

symbol $\binom{n}{\delta}$, because it affects only the two-holed sphere $L_{2,a}$.

Thus $\Delta_a(\binom{n}{\delta} q)) = \binom{n}{\delta+n} q)$, Next we shall consider not the action

of Δ_b, but that of $\Delta_a \Delta_b^{-1} \Delta_a$. This transformation carries a to

b^{-1} and b to a (see §4a)). Now $|n|$ is always the irreducible

number of intersections of the curve system with a, but δ is

only the irreducible number of intersections of the system with b

in case $|n| > q$, i.e. when there are no self-connections of the

boundary in $L_{3,a}$. If we now assume that q is not greater than $|n|$

and $|\delta|$, then $|\delta|$ is the number of intersections of the system

with b, and hence under the transformation $\Delta_a \Delta_b^{-1} \Delta_a$ it is the

number of intersections with a, and hence equal to $|n'|$, where

$\Delta_a \Delta_b^{-1} \Delta_a (\binom{n}{\delta} q)) = \binom{n'}{\delta'} q)$. $|n|$ is the number of intersections of the

system with a, hence under the mapping it becomes the number of

intersections with b, and hence equal to $|\delta'|$, because

$|n'| = |\delta| > q$ by hypothesis. The signs of δ and n are easy

to determine when one recalls that $\Delta_a \Delta_b^{-1} \Delta_a$ can be regarded as a

rotation of the system a,b. It follows from this, as with the

closed torus, that $n' = -\delta$, $\delta' = n$. By continued application of

the operations Δ_a and Δ_a^{-1}, and the operation $\Delta_a \Delta_b^{-1} \Delta_a$, we can

convert the symbol $\binom{n}{\delta}q)$ into $\binom{n'}{\delta'}q)$, where $|\delta'|$ is now $< q$.
As long as q is not greater than $|n|$ or $|\delta|$, the euclidean
algorithm holds. Thus we now consider the case $|n| > q > |\delta|$.
If $\Delta_a \Delta_b^{-1} \Delta_a (\binom{n}{\delta}q)) = \binom{n'}{\delta'}q)$ again, then $|n'| = |\delta|$ still. $|n|$
is the number of intersections the transformed system has with b,
but since $|n'| < q$ now, this number is also equal to
$|\delta'| + q - |n'|$, because $q - |n'|$ is the number of self-connections
of the boundary in $L_{3,a}$ in the transformed system. Thus
$|\delta'| = |n| + |\delta| - q$. The sign is also not difficult to determine,
and it turns out that n' has the opposite sign to δ, and δ' has
the sign of n. However, $|n|$ becomes $< q$ by these new operations,
and with a suitable number of operations Δ_a or Δ_a^{-1} we also get
$|\delta| < |n| < q$, so that we now have the case $q > |n| > |\delta|$. We
again denote $\Delta_a \Delta_b^{-1} \Delta_a (\binom{n}{\delta}q))$ by $\binom{n'}{\delta'}q)$. The number of intersections
with b is now $|\delta| + q - |n|$, thus $|n'| = |\delta| + q - |n|$. The
number of self-connections of the boundary in $L_{3,a}$ for the trans-
formed system is $q - |n'|$, hence equal to $|n| - |\delta|$. Thus
$|\delta'| = |n| - (|n|-|\delta|) = |\delta|$. The signs are as above.

It is now easy to see that there are only 12 _different_ curve
systems which satisfy the condition $q > |n| > |\delta|$ and which are
transformable into each other. Because in this case we have (see Fig. 19)
$|n| - |\delta|$ connections of two points on r which cut a once and
b not at all and which are homotopic to each other, $|\delta|$ connections
which cut a and b once and which are homotopic, and finally

q - |n| connections which cut only b, one time (the self-

connections of r in $L_{3,a}$), and which are also homotopic to each other.

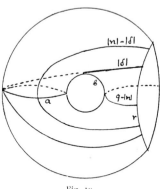

Fig. 19.

The curve systems therefore divide into 3 <u>divisions</u> and, when $(\frac{n'}{\delta}, q)$

is a symbol transformable into $(\frac{n}{\delta} q)$ which satisfies the same

inequalities, |n'| - |δ'| and |δ'| must each be equal to one of

the 3 numbers for the 3 divisions. But choice of the two divisions

and the evaluation of |n'| - |δ'| and |δ'| determines |n'| and

|δ'|. Thus we have altogether 6 different possibilities for the

pair |n'|, |δ'|, and since double application of the mapping

$\Delta_a \Delta_b^{-1} \Delta_a$ simultaneously changes the signs of n and δ, we have the

12 different possibilities claimed. At the same time, we see that

these can be reached by application of $\Delta_a \Delta_b^{-1} \Delta_a$ and Δ_a, so that we

have the result that Δ_a and $\Delta_a \Delta_b^{-1} \Delta_a$ generate all mappings of the perforated torus. This is only a confirmation of the earlier result, since the generation from Δ_a and Δ_b immediately yields the generation by Δ_a and $\Delta_a \Delta_b^{-1} \Delta_a$. At the same time, it follows that each curve system on the one-holed torus divides into 3 divisions, whose numbers can be determined by our recursion process (see the example).

An overview of the generation of the 12 different possibilities for n and δ, when $q > |n| > |\delta|$, is still lacking. We shall deal with this simultaneously with a general <u>geometric representation</u> <u>of the arithmetic field and the transformations of it induced by mappings.</u> However, we confine ourselves to the representation of transformations of $|n|$ and $|\delta|$. We construct an ordinary square lattice for these numbers (see Figs. 20 and 21) and denote the square with

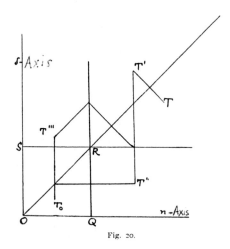

Fig. 20.

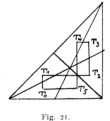

Fig. 21.

vertices $(0,0)$, $(q,0)$, (q,q), $(0,q)$ by OQRS. In the sector $q \leqslant |n|$, $q \leqslant |\delta|$ the ordinary operations of the euclidean algorithm,

$$\binom{|n|}{|\delta|} \rightarrow \binom{|\delta|}{|n|} \quad \text{and} \quad \binom{|n|}{|\delta|} \rightarrow \binom{|n|}{|\delta \pm n|} ,$$

then represent such transformations. By means of these operations, one gets from any point of the sector into the right-angled half strip $|n| \geqslant q > |\delta|$. In turn, one can get from the latter, by means of the operation,

$$\binom{|n|}{|\delta|} \rightarrow \binom{|\delta|}{|n|-q+|\delta|}$$

into the oblique-angled half-strip $q > |n| < |\delta|$. Finally, when $n \neq 0$, one can get from the latter into the triangle OQR, $q > |n| > |\delta|$ by suitable repetition of the operation $\binom{|n|}{|\delta|} \rightarrow \binom{|n|}{|\delta|-|n|}$.

This triangle is subdivided into 6 triangles by the medians. By suitable repetition of the two operations

$$\binom{|n|}{|\delta|} \rightarrow \binom{|\delta|+q-|n|}{|\delta|} \quad \text{and} \quad \binom{|n|}{|\delta|} \rightarrow \binom{|n|}{|n|-|\delta|} ,$$

a lattice point in one of the 6 triangles can be carried to any other. These 6 triangles correspond to the 6 different permutations of the 3 divisions of curves, or to the 6 different distributions of divisions among the divisions of curves which cut a_i and b_i, only a_i, or only b_i. On the medians the two division numbers are equal. On the sides of the big triangle a division number equals 0, hence at the vertices two division numbers equal 0. As we saw, each lattice point can be brought into one of the 6 triangles by an operation, provided it neither lies on $(0,|\delta|)$ nor is brought there by an operation. But $(0,|\delta|)$ in the oblique-angled strip results from $(|n|,0)$ in the right-angled strip $|n| > q > |\delta|$. This lattice point results from the point with coordinates $|n|$ and $|\delta|$ for $(n,\delta) > q$. All lattice points for which the greatest common divisor of n and δ is smaller than q can therefore be sent into one of the 6 subtriangles by our operations. All curve systems for which $(n,\delta) > q$ are transformable into curve systems consisting of $(n,\delta)-q$ closed curves parallel to b_i and q self-connections of the boundary in L_3,b, because such a system is represented by $\binom{n}{0}\,q)$ for $|n| > q$. Thus a subtriangle, together with the half-line $\delta = 0$, $|n| > q$, represents a fundamental domain on the arithmetic field or for the curve systems under transformations with movable boundary. The fundamental domain for the closed torus is given by the half line $\delta = 0$, because one can reduce each $\binom{n}{\delta}$ to $\binom{(n,\delta)}{\sigma}$. We also remark that two different lattice points in the same subtriangle give two different triples of division numbers.

Examples: 1) $\binom{25}{11}\,7)$, $|n|$ and $|\delta| > q$.

One obtains $\begin{pmatrix} |\delta| \\ |n| \end{pmatrix},\,q)$, i.e. $\binom{11}{25}\,7)$ by $\Delta_a\Delta_b^{-1}\Delta_a$, and $\binom{11}{3}\,7)$ by

Δ_a^{-2}.   Now we are in the region $n > q > \delta$ and we obtain

$$\frac{|\delta|}{|n|-q+|\delta|} \quad , \text{ i.e. } \quad \binom{3}{7} \ 7) \quad \text{by} \quad \Delta_a\Delta_b^{-1}\Delta_a, \text{ and } \quad \binom{3}{1} \ 7) \quad \text{by} \quad \Delta_a^{-2}.$$

This brings us into a subtriangle and hence we obtain the division

numbers : $|\delta| = 1$, $|n|-|\delta| = 2$, $q-|n| = 4$.   We confirm this by

direct consideration of the curve system $\binom{25}{11}$ 7) : we number the

points on a_1 and a_2 from 1 to 25.   The first 7 points on

a_1 and a_2 have connecting lines in $L_{3,a}$ with   r, the rest are

connected in $L_{3,a}$ to a_2 and a_1 respectively.   The passage

from a_1 to a_2 in $L_{2,a}$ occurs by addition of   11 modulo 25, the

passage from a_1 to a_2 in $L_{3,a}$ by connection of like-numbered

points.   We go from   r   to the point   1   on a_1 and then obtain

the series:

1   (on a_1), 12, 23, 9, 20, 6 on a_2 and, since this point is

connected to   r   in $L_{3,a}$,   1   on a_1 is connected to   r   by a   4-tuple

circling of   b, 2-tuple circling of   a.

Leaving   2   on   r, we obtain a parallel curve: 2, 13, 24, 10,

21, 7 ; but leaving   3   on   r   yields the series   3, 14, 25, 11, 22,

8, 19, 5, and thus a   6-tuple circling of   b, 3-tuple circling of   a ;

leaving   4, 5, 6, 7   on   r, we obtain the four similarly connected

series:

4, 15, 1

5, 16, 2

6, 17, 3

7, 18, 4, each of which is a   1-tuple circling of   b   and   a

1-tuple circling of a.   Thus in fact we obtain 3 divisions with

2, 1, 4 elements respectively, as we have already found by our

arithmetic process above.   We also remark that the fact, resulting

here from topological considerations, that the direct process always

leads to only 3 divisions with series of different length, is a

purely arithmetic theorem.   The purely arithmetic proof could

scarcely be carried out otherwise than by proving that the division

numbers are preserved by our "process".

2) $\binom{23}{11} 7) \rightarrow \binom{11}{23} 7) \rightarrow \binom{11}{1} 7) \rightarrow \binom{1}{5} 7) \rightarrow \binom{1}{0} 7).$

Thus we have only 2 divisions, with one member and 6 members

respectively.   To check, we construct the series

    1, 12, 23, 11, 22, 10, 21, 9, 20, 8, 19, 7 ;

    2, 13, 1

    3, 14, 2

    4, 15, 3

    5, 16, 4

    6, 17, 5

    7, 18, 6,

which in fact divide into 2 different divisions with one member and

6 members respectively.

Finally, we remark that a common divisor of n, δ and q

is not lost by the process and it also appears in the division

numbers.   Consequently, the symbol $\binom{\rho n}{\rho \delta} \rho q)$ represents the curves

with the symbol $\binom{n}{\delta} q)$, each provided with $\rho-1$ parallel curves.

When closed curves which are not parallel to r appear, then there can be only 2 because if we cut R_1^1 along such a curve, say c, the result is an $L_{3,c}$. Curves on this $L_{3,c}$ which do not meet two boundaries and which are not parallel to the third boundary are either parallel to these boundaries, and hence homotopic to c on R_1, or else self-connections of the third boundary. The symbols corresponding to such curve systems are always of the kind for which either n = 0 or $(n,\delta) > q$.

When no closed curves appear, then all curves are bounded by 2q points on r. They divide, as we know, into three pairs of divisions. We see from Fig. 19 that each pair of corresponding points in one division separates each pair of corresponding points from another division on r. The pairs from the same division do not separate each other. This corresponds precisely to the result[1] for polyhedra with p = 1 and one vertex, with no interior vertices, in an earlier work. The one vertex corresponds to the boundary here. This result can be derived quite easily without use of the symbols for the curve system, i.e. without passing through the arithmetic field.

c) <u>One holed torus</u> R_1^1 <u>with fixed boundary</u>

On R_1^1 the mapping $T^4 = (\Delta_a \Delta_b^{-1} \Delta_a)^4$ with movable boundary belongs to the class of the identity. It is easy to see that <u>with fixed</u>

---

[1] See Acta math. 67, p. 165 ff.

boundary T^4 represents a _twist along the boundary._ In fact a

mapping in the class of T transforms the connecting line A_1BA_2

(see Figs. 22 and 23) into A_1CA_2, a mapping of the class T^2 trans-

forms A_1BA_2 into $A_1D_1E_2BE_1D_2A_2$, and finally a mapping of the

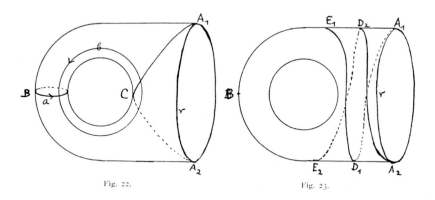

Fig. 22.   Fig. 23.

class T^4 transforms A_1BA_2 into $A_1D_1E_1BE_2D_2A_2$, which results from

A_1BA_2 by the twist Δ_r along the boundary. Thus $T^4\Delta_r^{-1}$ carries

the curves a and b, as well as the connecting line A_1BA_2, into

themselves. If we cut R_1^1 along A_1BA_2, the result is an L_2 for

which the only mappings are twists along the boundary, i.e. twists

along b. But these twists do _not_ map a onto itself, hence

$T^4\Delta_r^{-1}$ belongs to the identity class of mappings of R_1^1 onto itself.

Thus T^4 belongs to the class of Δ_r and represents a twist along

the boundary.

The mapping class group for R_1^1 with fixed boundary is generated by the mappings Σ and T with the relation $\Sigma^3 T^2 = 1$ (see §5d). This relation holds with a fixed boundary, because $\Sigma^3 T^2 = 1$ expresses the fact that the twist Δ_b goes to the twist Δ_a^{-1} by a mapping (namely $\Delta_a \Delta_b^{-1} \Delta_a$). This must also hold with a fixed boundary, because a twist always goes to a twist under a mapping, with fixed or movable boundary. Thus Δ_b cannot go to the twist Δ_a^{-1} connected with a twist along the boundary. On the other hand, the second relation $T^4 = 1$ does not hold, because $A_1 BA_2$ and $T^4(A_1 BA_2)$ are not homotopic with fixed boundary, as one sees immediately by consideration of the universal covering surface of R_1^1. It then follows from the normal form of elements generated by Σ and T on the basis of the relation $\Sigma^3 T^2 = 1$ that this relation alone defines the mapping class group of R_1^1, and hence this group is identical to the group of the trefoil knot.

§7.

Arithmetic field and curve systems on the four-holed sphere

a) Systems of closed curves on the L_4 (Fig. 24)

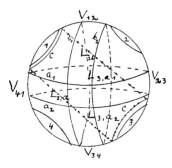

Fig. 24.

We divide the L_4 by two parallel curves a_1 and a_2 into two three-holed spheres L_{3,a_1} and L_{3,a_2} and a two-holed sphere $L_{2,a}$. Since the curves of the system are to be closed, all curves on the three-holed spheres which are not parallel to the boundaries or to a_1 and a_2 are self-connections of the boundaries a_1 and a_2 respectively. We join the four boundaries of the L_4 with normal connections V_{12}, V_{23}, V_{34}, V_{41}. V_{23} and V_{41} cut each of a_1 and a_2 at one point. The curves of the system have, say, $2n$ points in common with each of a_1 and a_2. We then place n of the points on a_1 and a_2 on one of the two parts between V_{23} and V_{41} and fix connections in L_{3,a_1} and L_{3,a_2} between any two such points, cutting V_{12} resp. V_{34} once and V_{23} and V_{41} not at all. Each system of closed curves on L_4 which includes no curves parallel to the boundaries or a_1 and a_2 is homotopically transformable to these connections inside L_{3,a_1} and L_{3,a_2} <u>with movable boundaries.</u> Only the connections in the $L_{2,a}$ remain, and these can be determined by a number δ as before. Thus the curves of the system are given by the symbol $\binom{n}{\delta}$, where δ is an integer, positive, negative or zero, but n is a natural number. The symbol $\binom{0}{\delta}$ is associated with δ curves parallel to a_1, oriented according to the sign of δ. It is not possible to give an orientation to general curves without something extra; we come back to this later. The symbol $\binom{n}{\delta}$ will represent all systems of closed curves on the L_4 which contain no curves parallel to the boundaries. And in fact n and δ

are independent of homotopies with movable (but not exchangeable) boundaries. One can prove this by direct topological considerations, though perhaps more intricately. It is more convenient to again use the representation by the fundamental group of L_4, which is the free group generated by 3 of the operations r_1, r_2, r_3, r_4 which represent circuits around the 4 boundary curves. In fact it is easy to obtain the representation of the curve system in the fundamental group from our symbol. However, we shall first simplify this problem by supposing that our curve system consists of a single curve. We shall see later that, just as on the torus, the symbol $\binom{n}{\delta}$ represents a single curve if and only if $(n, \delta) = 1$, and that the general curve system consists of (n, δ) parallel curves. Thus, for our proof of the claimed homotopy invariance of $\binom{n}{\delta}$, it suffices to consider the case $(n, \delta) = 1$. Also, we first assume $0 < \delta < n$. It then follows by consideration of the path in the dissected surface (cf. p. 275 above and Fig. 25, where $\binom{n}{\delta} = \binom{5}{2}$), that the curve $\binom{n}{\delta}$

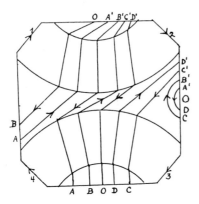

Fig. 25.

is represented by a 2n-termed expression in the generators $r_1^{\pm 1}$, $r_3^{\pm 1}$, $r_4^{\pm 1}$. The generator $r_3^{\pm 1}$ appears δ times in this expression, the generator $r_4^{\pm 1}$ appears $n-\delta$ times, and $r_1^{\pm 1}$ appears n times. Since a conjugation makes only a cyclic permutation of the terms in the expression (except that successive terms $\theta\theta^{-1}$ which occur under cyclic permutation may be omitted), because of the freeness of the generators, n and δ are uniquely determined by this expression or any of its conjugates, i.e. in this case n and δ remain unchanged under homotopies. However, if $|\delta| < n$ but $0 > \delta$, then the same consideration applies when we introduce the generator r_2 in place of r_3 and represent the fundamental group as the free group on the generators r_1, r_2, r_3.

The case $|\delta| > n$ now remains. We reduce this to the preceding by a "coordinate transformation", which is basic to our understanding of the L_4 : namely, instead of determining the curve system by the n points of intersection on the parallels a_1 and a_2 and the measure of twist δ, we determine it by the n' points of inter-section with the curves b_1 and b_2 (see Fig. 24) and the measure of twist δ' in the $L_{2,b}$ between b_1 and b_2. We place b_1 and b_2 so that all self-connections of the a_i lie in the $L_{2,b}$. Then we have exactly $|\delta|$ self-connections of the b_i in the L_{3,b_i}. Thus $n' = |\delta|$, and all these self-connections lie in the $L_{2,a}$. By the reverse argument we obtain $n = |\delta'|$. Finally, we give δ' the sign of δ by a suitable determination of the positive side of a_1.

The coordinate transformation then reads : $n' = |\delta|$, $\delta' = \frac{|\delta|}{\delta}n$.
The contrast with the analogous transformation on the torus R_1
is clear : on the L_4 we are dealing only with a coordinate transformation, whereas on R_1 this can also be regarded as a mapping.
This is out of the question for the L_4 under the assumption of
non-exchangeable boundaries, because the b_i separate pairs of
boundaries different from those separated by the a_i. When
boundaries can be exchanged, the coordinate transformation corres-
ponds to a quarter turn. The case $|\delta| > n$ has now been reduced
to the case $|\delta| < n$.

Only the case $n = |\delta| = 1$ now remains. A symbol with other
n or δ cannot go to $\binom{1}{\pm 1}$ under homotopic transformation, because
we have proved the invariance of n and $|\delta|$ for the other symbols.
But $\binom{n}{\delta}$ also cannot go to $\binom{n}{-\delta}$. Because if we transform both
curves by a twist in $L_{2,a}$, then they go to curves with the symbols
$\binom{n}{\delta+2n}$ and $\binom{n}{-\delta+2n}$, which, since the numbers $|\delta+2n|$ and $|-\delta+2n|$
are different for $\delta \neq 0$, represent two non-homotopic curves. Thus
$\binom{n}{\delta}$ also does not go to $\binom{n}{-\delta}$ by a homotopy, and we have now
proved the invariance of $\binom{n}{\delta}$ under homotopy in general. Here it
is equally valid to consider the boundaries as fixed or movable,
since we are dealing purely with closed curves.

b) Mappings of the four-holed sphere onto itself and the action
on the arithmetic field. In §3 we have considered special mappings
of L_4 onto itself, the twists of $L_{2,a}$ and $L_{2,b}$. It follows

from the early results and the above development that the first

twist carries n to n, $|\delta|$ to $|\delta+2kn|$. and the second carries

n to $|n+2k'\delta|$, $|\delta|$ to $|\delta|$. Thus we again have (at first only

for the absolute values) a euclidean algorithm, but of such a

kind that we can only add or subtract an even multiple of one number

to or from the other. These operations constitute a (non-invariant)

subgroup of the operations of the general euclidean algorithm. We

shall call them even-euclidean operations. If $|\delta| > n > 0$, then

we can determine k so that $|\delta+2kn| \leqslant n$, but if $n > |\delta| > 0$ we

can determine k' so that $|n+2k'\delta| < |\delta|$. The process breaks off

in the following three cases : 1) n = 0, 2) $\delta = 0$, 3) $n = |\delta|$.

In the first case we obtain $|\delta|$ curves parallel to a_1, in the second

n curves parallel to b_1, in the third case we have $\delta = \pm n$.

However, the symbol $(\begin{smallmatrix} n \\ -n \end{smallmatrix})$ goes to $(\begin{smallmatrix} n \\ n \end{smallmatrix})$ by addition of 2n to -n.

That is n curves parallel to a certain curve, say c, separating

r_2 and r_4 from r_1 and r_3 (see Fig. 24). Thus we have the

Result : each system of disjoint simple closed curves on the four-holed

sphere can be mapped, by means of the twists along a and b,

onto a number of curves parallel to one of three curves which separate

two of the boundaries from the others. This number is equal to

(n,δ) when the curve system is represented by $(\begin{smallmatrix} n \\ \delta \end{smallmatrix})$. Here, curves

parallel to the boundaries are ignored. In this way we have derived

the fact used above, that when $(n,\delta) = 1$ the curve system consists

of only one curve. In doing so we have not used the theorem on the homotopy invariance of the symbol.

Now a mapping which sends, say, a_1 and a_2 to themselves can be generated on L_{3,a_1} and L_{3,a_2} by homotopy (with movable boundaries). In the $L_{2,a}$ it generates a twist, which corresponds to the transformation $\binom{n}{\delta} \to \binom{n}{\delta+kn}$. A mapping Φ of the L_4 which sends two curves α_1 and α_2 to a_1 and a_2 can then be transformed by a suitable $\bar{\Phi}$, composed of twists in $L_{2,a}$ and $L_{2,b}$, to send α_1 and α_2 to curves parallel to a_1, b_1 or c. But they must be curves parallel to a_1, because a_1 and α_1 separate the same pairs of boundaries by hypothesis. Thus $E\bar{\Phi}(\alpha_1) = a_1$, $E'\bar{\Phi}^{-1}(a_1) = a_1$. Consequently, $\Phi\bar{\Phi}^{-1}$ equals a twist in $L_{2,a}$, up to homotopy. Thus, representatives of each mapping class of the four-holed sphere can be generated by twists in $L_{2,a}$ and $L_{2,b}$, which correspond to the even euclidean operations. However, if an even euclidean operation is not the identity transformation, i.e. if it does not carry all symbols to themselves, then it represents a non-homotopy transformation, since, as we have seen, homotopies carry all symbols to themselves. On the other hand, a non-homotopy transformation sends at least one of the symbols $\binom{1}{0}$ or $\binom{0}{1}$ to another symbol, and therefore corresponds to a non-identity transformation in the group of even euclidean operations. We therefore have: the even euclidean transformations generate a group isomorphic to the mapping class group of the four-holed sphere.[1]

---

[1] W. MAGNUS informs me that it follows from this, by means of the result of H. FRASCH, Math. Ann. 108, p. 245, that the mapping class group for the L_4 is the free group on two generators.

c) It is easy to see which symbols represent curve systems which can be mapped onto a particular one of the three systems of parallel curves. Because, <u>the even euclidean operations preserve the parity of</u> n <u>and</u> δ. We can assume that n and δ are not both even, otherwise the greatest power of 2, say 2^ℓ, which divides them both, could be taken out, and for that reason $\binom{n}{\delta}$ would represent a curve system consisting of 2^ℓ parallels to the system $\left(\begin{array}{c} \frac{n}{2^\ell} \\ \frac{\delta}{2^\ell} \end{array} \right)$, where $\frac{n}{2^\ell}$ and $\frac{\delta}{2^\ell}$ are not both even. We then have three cases : 1) n and δ odd, $\binom{n}{\delta}$ represents a curve system which can be mapped onto (n,δ) curves parallel to c.

2) n even, δ odd, $\binom{n}{\delta}$ represents a curve system which can be mapped onto (n,δ) curves parallel to a. 3) n odd, δ even, the curve system can be mapped onto (n,δ) curves parallel to b. In this way we have derived <u>the invariants of a curve system under mappings, from the symbols.</u>

d) An orientation of the system $\binom{n}{\delta}$ induces one on r_1, chosen so that the piece of L_4 between r_1 and a curve of the system receives the same indicatrix from both sides. A mapping sends r_1 to itself with orientation preserved, and hence the orientation of the system $\binom{n}{\delta}$ is carried over to its image by means of r_1, and hence onto the curve a, b, or c onto which $\binom{n}{\delta}$ can be mapped. It is easy (see Fig. 26) to carry over orientation

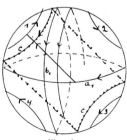

Fig. 26.

from r_1 to a, b, c on the one hand and — by the first self-connection of a in L_{3, a_1} — to $\binom{n}{\delta}$ on the other.

We shall now take a particular orientation of r_1 as positive, and thereby also take a particular orientation of a_1, b_1, c and $\binom{n}{\delta}$ as positive. We shall take $\binom{n}{\delta}$ to be the positively oriented curve, or positively directed parallel system in case $(n, \delta) > 1$, which was also denoted by $\binom{n}{\delta}$ previously. The oppositely directed curves will from now on be denoted by the symbol $\binom{-n}{\delta}$. Our coordinate transformation $(a \rightarrow b)$ is now seen as:

$$\bar{n} = \frac{|n|}{n} |\delta|, \quad \bar{\delta} = \frac{|\delta|}{\delta} |n|.$$

In this way the upper symbol of the pair, n, $\bar{n}$, etc. always retains the same sign under the operations of coordinate transformation and the euclidean algorithm. Now we shall also, as we did earlier with the symbols for curve systems on the torus, give the symbol δ a different geometric meaning according as n is positive or negative, and in fact change of orientation of the curves shall also change the circling direction for δ or, in other words, when $\binom{n}{\delta}$ represents a curve system, $\binom{-n}{-\delta}$ shall represent the same curve system with the opposite orientation. This convention is necessary in order to represent one mapping by the same transformation formula $\delta' = \delta + kn$ for arbitrary orientation (without this convention k would have to change its sign with the sign of n for the same mapping.)

Examples :   1)

$$c \equiv \binom{1}{1}, \text{ by twist in } L_{2,a} \rightarrow \binom{1}{-1},$$

by coordinate transformation $a \rightarrow b : \rightarrow \binom{1}{-1}$,

by coordinate transformation $b \rightarrow a : \binom{1}{1}$

i.e. <u>a twist in</u> $L_{2,a}$ <u>and a twist in</u> $L_{2,b}$ yields a mapping which

sends $c \equiv \binom{1}{1}$ to itself, hence it is a <u>twist along</u> c.

2)

$$c^{-1} \equiv \binom{-1}{-1} \rightarrow \binom{-1}{1} \text{ by coordinate transformation } a \rightarrow b :$$

$$\rightarrow \binom{-1}{1} \rightarrow \binom{-1}{-1} \text{ by coordinate transformation } b \rightarrow a :$$

$$\rightarrow \binom{-1}{-1}.$$

The result of this section is : <u>there is a one-to-one correspondence</u>
<u>between the systems of closed, like-oriented curves, disregarding</u>
<u>curves parallel to the boundaries, and the symbols</u> $\binom{n}{\delta}$, <u>where</u> n
<u>and</u> δ <u>are positive, negative or zero numbers, but not both zero.</u>

e) <u>Geometric representation of the symbols</u> $\binom{n}{\delta}$

Once again we shall consider the original symbol $\binom{n}{\delta}$, where
n is a positive number.   The set of 2n symbols $S_1, \ldots, S_n$,
$S_{n+1}, \ldots, S_{2n}$ corresponding to the points on a_1 and a_2 is mapped
onto itself by $S_i \rightarrow S_{\bar{i}*}$, where $\bar{i}* \equiv i + \delta$ mod 2n, by a twist in
$L_{2,a}$.   The self-connections in L_{3,a_2} send $S_{\bar{i}*}$ to $S_{\bar{i}}$, where

$\bar{i} = 2n + 1 - \bar{i}*$, hence $\bar{i} \equiv 1 - i - \delta \bmod 2n$; then another twist

in $L_{2,a}$ sends $\bar{i}$ to $i'*$, where $i'* \equiv \bar{i} - \delta \equiv 1 - i - 2\delta \bmod 2n$,

and then a self-connection in L_{3,a_1} sends it to

$i' = 2n + 1 - i'* \equiv i + 2\delta \bmod 2n$.

Thus the symbol $\binom{n}{\delta}$ yields a substitution

$$i' \equiv i + 2\delta \bmod 2n$$

under a single circling of b. This substitution is cyclic, and

hence all cycles into which it decomposes have the same number of

terms, and in fact there are exactly (n,δ) such cycles. This

corresponds to the above result that the curves of a system are all

parallel, and (n,δ) in number.

The connection between the symbol $\binom{n}{\delta}$ and the above substitution

leads to a graphical representation of the curve system (see Fig. 27):

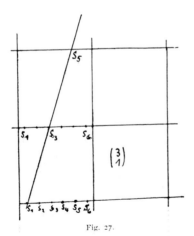

Fig. 27.

we use a square net and place the points $S_1, \ldots, S_{2n}$ on all horizontal sides in the same order, at distance ε from each other, and with the points S_1 and S_{2n} at distance $\frac{\varepsilon}{2}$ from the vertices. Then the curve system represented by $\binom{n}{\delta}$ is also represented by all lines which connect a point S_i with abscissa x on the lower side of a horizontal strip with a point with abscissa $x + 2\delta\varepsilon$ on the upper side of the strip. Lines which are homologous in the net, i.e. differing only by parallel displacement of the net, are considered to be the same, i.e. lines are considered only "modulo the net". The lines yield the corresponding substitutions by their intersections with successive horizontal lines of the net. In place of the square net we can use a torus to represent the curve system, since the universal covering of the torus is the square net. However, it is obviously impossible to regard the square net as universal covering of the four-holed sphere.

As we have just seen, the curve system on L_4 corresponds to a substitution of $2n$ symbols, generated by a reflection I, namely the index substitution $i' = 2n + 1 - i$, and a cyclic shift $Z : i' \equiv i + \delta \bmod 2n$. The substitution $I^{-1}Z^{-1}IZ$ is again a shift : $i' = i + 2\delta \bmod 2n$. This is merely the simplest case of a very frequent phenomenon in topology. Let us consider, e.g., the double torus which results from joining two one-holed tori $R_{1,1}^1$ and $R_{1,2}^1$ along the boundary r. As we saw in §6b), the curve systems on $R_{1,1}^1$ and $R_{1,2}^1$ map onto curves in three divisions.

These divide the $2n$ intersection points on r into three pairs of divisions. Suppose we have the intersection points

$$S_1 \cdots S_{n_1} S_{n_1+1} \cdots S_{n_1+n_2} S_{n_1+n_2+1} \cdots S_{n_1+n_2+n_3=n}$$

$$S_{n+n_1} \cdots S_{n+1} S_{n+n_1+n_2} \cdots S_{n+n_1+1} S_{2n} \cdots S_{n+n_1+n_2+1}$$

for $R^1_{1,1}$. Here the points are ordered on r according to the natural order of indices 1 to $2n$. In the order written above, the endpoints associated by connections on $R^1_{1,1}$ stand in the same column. The resulting correspondence is a permutation of the naturally ordered $2n$ symbols of order 2, call it I. We have a similar substitution I' of the $2n$ symbols corresponding to the curves on $R^1_{1,2}$ with other division members, say n'_1, n'_2, n'_3. The curve system on the double torus results from a twist Z along r, i.e. a substitution $i' \equiv i + \delta$ (mod $2n$), and this curve system therefore corresponds to the substitution $ZIZ^{-1}I'^{-1}$. For other topological figures we have quite differently constructed substitutions I of second order, e.g. for curve systems on the six-holed sphere, where the three division pairs follow each other in the order $n_1 n'_1 n_2 n'_2 n_3 n'_3$, as Fig. 28 shows. In the simplest case, which we

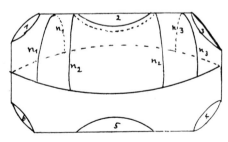

Fig. 28.

have considered above, where only one pair of divisions is present and (quite unimportantly, however) $I = I'$, $ZIZ^{-1}I^{-1}$ was again a cyclic substitution and the euclidean algorithm gave us the essential properties. In other cases one at present knows only very unclear processes based on topological considerations. Even a simple transformation corresponding to the above "coordinate transformation", the exchange of n and δ, is lacking. Perhaps the geometric representation of the simplest case by a square net may be carried over to more complicated cases by regular subdivisions of the non-euclidean plane.

f) <u>Curve systems on the four-holed sphere with endpoints on</u>
   <u>one boundary</u>

1) We shall proceed differently from our treatment of the one-holed torus by first making a direct derivation, by adding some topological considerations to the previous results. If we cut the L_4 along a self-connection σ of the boundary r_1 which is not contractible to r_1 (see Fig. 29), the result is an $L_{3,\sigma}$ and an $L_{2,\sigma}$.

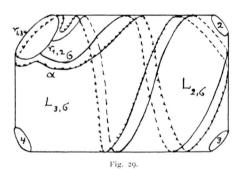

Fig. 29.

On the $L_{2,\sigma}$ there is a second boundary of the L_4, say r_2. The boundaries r_3 and r_4 are then the other two boundaries of the $L_{3,\sigma}$. The boundary of $L_{2,\sigma}$ other than r_2 is a curve consisting of the self-connection σ and a piece $r_{1,2}$ of the boundary r_1 between the endpoints of σ. Similarly, the boundary of the $L_{3,\sigma}$ other than r_3 and r_4 is a curve made up of σ and the other piece $r_{1,34}$ of r_1 between the endpoints of σ.

We now consider a curve α parallel to this latter boundary in $L_{3,\sigma}$. This is a closed curve on the L_4 which separates r_1 and r_2 from r_3 and r_4. It then follows from previous results that it can be mapped onto the curve a_1 by a mapping of L_4 onto itself with movable, but not exchangeable, boundaries. Under this mapping σ goes to a self-connection $\bar{s}$ of r_1 which lies in L_{3,a_1} and which, by results proved earlier (§3), is homotopic with movable boundaries to a certain self-connection s of the boundary r_1 in L_{3,a_1}, and hence also on L_4.

If we have several (disjoint) self-connections σ_i of r_1 which separate r_2 from r_3 and r_4, then we cut along the connection σ_0 for which the boundary piece $r_{1,34}$ of r_1, which belongs to $L_{3,\sigma}$, contains no endpoints of other such self-connections. A curve α parallel to $\sigma_0 r_{1,34}$ in $L_{3,\sigma}$ can again be carried to a_1 by a mapping. Now all these self-connections lie in L_{3,a_1} and they can be carried to fixed self-connections of r_1 by a homotopy

with movable boundaries.   We now consider the three-holed sphere
$L_{3,\sigma}$.   It contains any self-connections of r_1 which are not
homotopic to σ_0, because all those lying on the two-holed sphere
$L_{2,\sigma}$ (and which are not homotopic to a piece of r_1)   are homotopic
to σ_0.   The endpoints of every other connection are movable on
$r_{1,34}$ by a homotopy of the L_4, but naturally not across σ_0.
Thus we have here the case of the fixed boundary, and by the preceding
we obtain two different kinds of self-connection of r_1, not homotopic
to each other or to a boundary piece, and especially not homotopic
to σ_0, which differ from each other in the fact that one, say $\sigma_i{}'$,
encloses r_3 together with a piece of $r_{1,34}$, while the other,
say $\sigma_i{}''$, similarly encloses r_4 (see p.294 and Fig. 29a).

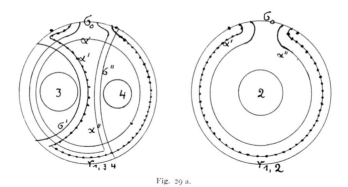

Fig. 29 a.

The mapping of α onto a_1 sends the first onto the three-holed
sphere L_{3,b_1}.   Because the parallel curve α', which is constructed

relative to σ_i' as α is above to σ_i, twice cuts α, or its image a_1 under the mapping $\alpha \to a_1$, as we see from the figure (α' is shown dotted in both figures of 29a), and therefore has (after the mapping $\alpha \to a_1$) the symbol $\binom{1}{\delta}$, and can be mapped by a twist in $L_{2,a}$ onto $\binom{1}{0}$, i.e. b_1, or $\binom{1}{1}$, i.e. c. But it can only be b_1 (i.e. $\binom{1}{0}$ and δ even) in case α' separates r_1 and r_4 from r_2 and r_3, as supposed. Similarly, the curve α'' parallel to σ_i'' is sent by the mapping to a curve with the symbol $\binom{1}{\delta}$. However, since α'' also has two points in common with α', α'' must go to a curve with the symbol $\binom{1}{\pm 1}$ in case α' is mapped onto b_1. Now, we can convert the curve $\binom{1}{-1} \equiv c_1{}^*$ into $\binom{1}{1} \equiv c_1$ by a twist in $L_{2,a}$ or $L_{2,b}$, but in the process either b_1 or a_1 goes to a non-homotopic curve. When the curves a_1 parallel to s and b_1 parallel to s' both go to homotopic curves then only a homotopic mapping is possible, and connections s'' parallel to c_1 are not convertible into parallels to $c_1{}^*$. The connections s'' lie either in the three-holed sphere L_{3,c_1} (containing r_1) determined by $c_1 \equiv \binom{1}{1}$, or in the three-holed sphere $L_{3,c_1{}^*}$ (containing r_1) determined by $c_1{}^* \equiv \binom{1}{-1}$. In summary we have:

Each system of self-connections of a boundary r_1 can be carried by a single mapping onto a system of fixed connections consisting of three divisions, which lie in $L_{3,a_1} L_{3,b_1} L_{3,c_1}$ or $L_{3,a_1} L_{3,b_1} L_{3,c_1{}^*}$.

It is clear that one can prescribe one of the three-holed spheres L_{3,c_1} or L_{3,c_1*}, but then one must admit L_{3,a_1*} as well as L_{3,a_1}, or else L_{3,b_1*} as well as L_{3,b_1}, according as one converts c_1 into c_1* by a twist along b_1 or a_1.

If our curve system consists, not only of self-connections of the boundary r_1, but also of closed curves, then we can convert all the closed curves into one of either a_1, b_1 or c_1, and the self-connections of r_1 all go thereby into the division belonging to a_1, b_1 or c_1. Thus in this case we have, as on the one-holed torus, only two divisions : the closed curves and one division of self-connections.

2) In 1) we have found a normal form for (unoriented) curve systems with endpoints on one boundary, and we shall now see how this normal form is expressed in symbols. As before, the symbols are introduced by dividing the L_4 by a_1 and a_2. Then the curves in L_{3,a_1}, and likewise in L_{3,a_2}, can be carried to fixed connections by homotopies with movable boundaries. All that now remain are the connections in $L_{2,a}$, which are determined by a twist number δ relative to a fixed connection, say the piece of b_1 between a_1 and a_2. The curve system is then determined up to homotopy by the number $2n$ of intersection points on a_1 and a_2, the number $2q$ of endpoints on r_1, and δ. We denote the curve system by $\binom{n}{\delta} q)$. In the case where the curve system is made up of ℓ closed curves

parallel to a_1 resp. b_1, resp. c_1 resp. c_1^* and m self-connections we have the symbols

$$\binom{0}{\ell}\ m)\ \text{resp.}\ \binom{m+\ell}{0}\ m)\ \text{resp.}\ \binom{m+\ell}{m+\ell}\ m)\ \text{resp.}\ \binom{m+\ell}{-m-\ell}\ m).$$

In the case where there are no closed curves, but instead m_a self-connections parallel to a_1, m_b self-connections parallel to b_1, and m_c self-connections parallel to c_1 resp. c_1^*, we have the symbols

$$\begin{pmatrix} m_a+m_c & & \\ & m_a+m_b+m_c \\ m_c & & \end{pmatrix}\ \text{resp.}\ \begin{pmatrix} m_a+m_c & & \\ & m_a+m_b+m_c \\ -m_c & & \end{pmatrix}.$$

We shall call this coordinate system the (a,b)-coordinate system (the second coordinate is decisive for the number δ). If we choose the corresponding (b,a)-coordinate system, then m_a and m_b are exchanged. We denote the symbol in the latter coordinate system by $\binom{n'}{\delta'}\ q)$. We see that when $m_b \neq 0$, $q > n$, and when $m_a \neq 0$, $q > n'$. However, if $m_a = m_b = 0$, then we introduce the (c,a)-coordinate system and obtain the symbol $\binom{0}{0}\ m_c)$ in case the self-connections are all parallel to c_1, which we can always arrange by twisting in $L_{2,a}$ or $L_{2,b}$. In the (c,a)-coordinate system we denote the symbol by $\binom{n''}{\delta''}\ q)$. We then have the result that for each curve system either n, n' or $n'' < q$ in a suitable coordinate system and after a suitable mapping. The (c,a)-coordinate system is necessary only when $n = |\delta| \geqslant q$, i.e. when $n = \delta \geqslant q$ after a possible twist in $L_{2,a}$. In this case the curve system is parallel to c_1 and contains $n-q$ closed curves.

We can avoid the topological considerations when we consider transformations of the symbols $\binom{n}{\delta} q$ as we previously (§6) considered the transformations of the analogous symbols for the one-holed torus. It is little more than a repetition, with the modification that now only the even euclidean operations appear: the reduction of the symbol $\binom{n}{\delta} q$ proceeds by first using the substitution $\bar{\delta} = \delta + 2kn$ to make $|\bar{\delta}| < n$. Then we go to the (b,a)-coordinate system and obtain the symbol $\binom{n_1}{\delta_1} q$ where $n_1 = |\bar{\delta}| < n$, and δ_1 is determined in exactly the same way as when $q = 0$ (no boundary points) in case $|\delta|$ and n are both $\geqslant q$. In this case $\delta_1 = \dfrac{\bar{\delta}}{|\bar{\delta}|} n$. By even euclidean operations we can therefore convert the symbol $\binom{n}{\delta} q$ into a symbol $\binom{n_{t-1}}{\delta_{t-1}} q$ for which $|\delta_{t-1}| < q$. The symbol represents a curve system, which results from the given one by mappings corresponding to even euclidean operations. However, if $n_{t-1} > q > |\delta_{t-1}|$ then

$$n_t = |\delta_{t-1}| \quad \text{and} \quad \delta_t = \frac{\delta_{t-1}}{|\delta_{t-1}|} (n_{t-1} - q + |\delta_{t-1}|)$$

(exactly as in the case of the one-holed torus) and in general we can proceed from $\binom{n_t}{\delta_t} q$, when $n \neq 0$, by the operation $\delta^* = \delta_t + 2kn_t$ to a symbol $\binom{n_t}{\delta^*} q$ in which n_t as well as $|\delta^*|$ is $< q$.

Here we have considered the symbols only as far as is necessary for handling the problem of the five-holed sphere. However, we also remark that in the (n,δ)-plane (for a particular coordinate system) a fundamental domain in this case is the isosceles right-angled triangle with the vertices $(0,0)$, (q,q), $(q,-q)$ together with the half-lines $n = \delta > q$; $n = 0$, $\delta > 0$ and $n > q$, $\delta = 0$.

g) L_4 with fixed boundaries

1) We saw above (Examples 1 & 3) that a twist in L_{2a} and a twist in L_{2b} generate a twist along c. Thus, with fixed boundaries, suitable twists along a, b and c yield a combination of twists along the four boundaries. We shall now determine this combination. We denote the twists along a, b, c and r_i by Δ_a, Δ_b, Δ_c and Δ_{r_i} and fix the sense of twist (see Figs. 30, 30a) as follows:

Δ_a replaces XX_0 by $XX_1X_2X_0$,

Δ_b replaces YY_0 by $YY_1Y_2Y_0$,

Δ_{r_1} replaces RR_0 by $RR_1R_2R_0$,

Δ_c^{-1} replaces ZZ_0 by $ZZ_1Z_2Z_3Z_0$

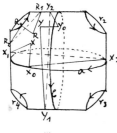

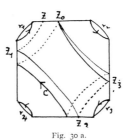

Fig. 30. Fig. 30 a.

The twists Δ_{r_i} commute with each other and with the twists Δ_a, Δ_b, Δ_c.

We now consider the line A_1A_3 connecting the first and third boundaries (see Fig. 31)

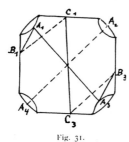

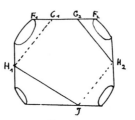

Fig. 31.   Fig. 31 a.

By our definition, $\Delta_b\Delta_a(A_1A_3)$ is homotopic to the connecting line $A_1B_1C_1C_3B_3A_3$. The latter also results from A_1A_3 by the mapping $\Delta_{r_1}\Delta_{r_2}$. Thus $\Delta_{r_1}^{-1}\Delta_{r_3}^{-1}\Delta_b\Delta_a$ maps the connection A_1A_3 onto itself. The mapping $\Delta_{r_2}^{-1}\Delta_{r_4}^{-1}\Delta_b\Delta_a$ similarly carries the connection A_2A_4 (see Fig. 31) to itself. However, since Δ_{r_1} and Δ_{r_3} do not alter the connection A_2A_4, and Δ_{r_2} and Δ_{r_4} do not alter the connection A_1A_3, $\Delta_{r_1}^{-1}\Delta_{r_2}^{-1}\Delta_{r_3}^{-1}\Delta_{r_4}^{-1}\Delta_b\Delta_a$ carries both connections A_1A_3 and A_2A_4 to themselves. If we cut the L_4 along these two connections, the result is an L_2 which is mapped onto itself by $\Delta_{r_1}^{-1}\Delta_{r_2}^{-1}\Delta_{r_3}^{-1}\Delta_{r_4}^{-1}\Delta_b\Delta_a$. Thus the latter is a mapping of the L_2, i.e. a twist along c, and hence equal to Δ_c^m. In order to find m, we consider the connection F_1F_2 (see Fig. 31a). We have

$$\Delta_{r_1}^{-1}\Delta_{r_2}^{-1}\Delta_{r_3}^{-1}\Delta_{r_4}^{-1}\Delta_b\Delta_a(F_1F_2) = \Delta_c^{-1}(F_1F_2),$$

as follows from the figure.   Thus $\Delta_{r_1}^{-1}\Delta_{r_2}^{-1}\Delta_{r_3}^{-1}\Delta_{r_4}^{-1}\Delta_c\Delta_b\Delta_a$ carries the

connections A_1A_3, A_2A_4 and F_1F_2 to themselves.   This mapping

therefore belongs to the class of the identity, and in the mapping

class group,

$$\Delta_c\Delta_b\Delta_a = \Delta_{r_1}\Delta_{r_2}\Delta_{r_3}\Delta_{r_4}.$$

2) We shall apply this result to the <u>two-holed torus with fixed</u>

<u>boundaries</u>.   By cutting along   a   (see Fig. 32) we see   from 1) that,

with suitable labelling,

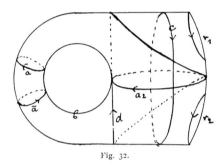

Fig. 32.

$$\Delta_{a_2}\Delta_c\Delta_d = \Delta_a\Delta_a\Delta_{r_1}\Delta_{r_2}.$$

Now Δ_c is generated by Δ_a and Δ_b, by §6c.   Thus Δ_{r_2} is

generated by $\Delta_{r_1}, \Delta_a, \Delta_{a_2}, \Delta_b$ and Δ_d : The twist along one boundary

is generated by twists along the other boundary and along the curves a, a_2, b and d which do not separate the torus.

3) Surfaces of higher genus with one boundary curve (R_1^p).

The surface is divided by p-1 separating curves $C_1, \ldots, C_{p-1}$
into an R_1^1 and p-1 R_2^1's.   (See Fig. 33.)   By §6c, the twist
along the boundary C_1 of the R_1^1 is generated by twists along

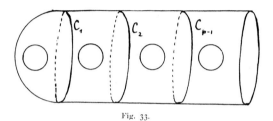

Fig. 33.

curves which are not separating on the (single) R_1^1, and hence
certainly not on the R_1^p.   By the result just proved in 2), the twist
along C_2 is generated by twists along C_1 and along curves which
are non-separating on the $R_{2,C_1C_2}^1$, and hence also on the R_1^p.
Thus the twist along C_2 may be generated by twists along curves
which do not separate the R_1^p.

Continuing in this way, we obtain the result: the twist along
the boundary of a surface R_1^p may be generated purely by twists
along non-separating curves of the surface.

4) <u>Twists along separating curves of an</u> R_0^p <u>or</u> R_1^p.

A separating curve on an R_0^p divides it into an R_1^q and an R_1^{p-q}, a separating curve on an R_1^p divides it into an R_1^q and an R_2^{p-q}. It follows from 3) that : <u>the twist along a separating curve of a surface with at most one boundary curve may be generated purely by twists along non-separating curves on the surface.</u>

Since the non-separating curves can all be sent to each other by mappings of the surface, i.e. they are all equivalent, and since mappings send twists to twists, we have the result: each mapping of a surface with at most one boundary curve generated by twists may be generated purely by mappings which are conjugate in the mapping class group.

§ 8

The five-holed sphere

a) <u>Coordinate systems.</u> On the L_4 we have three closed curves a, b and c which intersect pairwise in two points and form the basis for the determination of any curve system. Each closed curve of L_4 (which is not contractible nor homotopic to a boundary) may be mapped onto one of them. One thinks analogously of introducing ten closed curves on the L_5, each of which separates two of the boundaries from the other three. If we denote such a curve by $a_{ik} = a_{ki}$ $(i \neq k)$ in case it separates the boundaries r_i and r_k from the other boundaries, then the ten curves a_{ik} must satisfy the condition that a_{ik} and $a_{k\ell}$ intersect in two points, but a_{ik} and $a_{\ell m}$ ($\ell \neq i,k$; $m \neq i,k$) have no point in common. However, this is impossible because, as is easily seen, the latter condition will not be satisfied for at least one pair. This is because five points on the sphere cannot be connected with each other by segments which do not meet outside the five given points. We therefore content ourselves with five closed curves as "coordinate axes", say a_{12}, a_{23}, a_{34}, a_{45} and a_{51} (see Fig. 34).

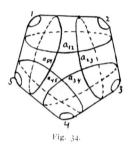

Fig. 34.

Each of these curves $a_{i,i+1}$ divides the surface into an $L_{4,i}$ and an $L_{3,i}$. The $L_{4,i}$ contains the two curves disjoint from $a_{i,i+1}$, namely $a_{i+2,i+3}$ and $a_{i+3,i+4}$. The indices here are always reduced modulo 5.

  b) Representation of a system of closed curves on the L_5.

We divide the surface into an $L_{4,i}$ and an $L_{3,i}$ by a curve $a_{i,i+1}$ with i = 4, say. In $L_{4,4}$ and $L_{3,4}$ we draw curves a'_{45} and a''_{45} parallel to a_{45}. Then the curve system on the $L_{4,4}$ is given by the number n_{12} of intersections with the curve a_{12}, a twist number δ_{12}, and the number n_{45} of boundary points on a'_{45} ; n_{45} is also the number of intersections of the system with

a_{45}. On the $L_{3,4}$, with boundaries a_{45}'', r_4 and r_5, the curve system consists, apart from possible parallels to a_{45}'', of self-connections of the boundary a_{45}'', and hence it is given by the number n_{45} (with movable boundary a_{45}''). There still remains the curve system on the $L_{2,4}$ with the boundaries a_{45}' and a_{45}'', which is determined by a twist number δ_{45}. The curve system is therefore fully determined by the pair of symbols

$$\begin{pmatrix} n_{12} & \\ & n_{45} \\ \delta_{12} & \end{pmatrix} \begin{pmatrix} n_{45} \\ \delta_{45} \end{pmatrix}.$$

We can similarly divide the surface by a_{12}. Then for the same curve system we have the symbol pair

$$\begin{pmatrix} n_{12} \\ \bar{\delta}_{12} \end{pmatrix} \begin{pmatrix} n_{45} & \\ & n_{12} \\ \bar{\delta}_{45} & \end{pmatrix}.$$

We shall denote this coordinate transformation, an exchange of a_{12} and a_{45}, by W_{14}. If, on the other hand, we carry out an exchange of coordinate systems on $L_{4,4}$, from (a_{12}, a_{23}) to (a_{23}, a_{12}) (see the exchange of a- and b-axes on the four-holed sphere in §7), then we obtain the pair

$$\begin{pmatrix} n_{23} & \\ & n_{45} \\ \delta_{23} & \end{pmatrix} \begin{pmatrix} n_{45} \\ \delta_{45}' \end{pmatrix}.$$

If, e.g., $n_{45} < n_{12}$ and $< |\delta_{12}|$, then

$$n_{23} = |\delta_{12}|, \quad \delta_{23} = \frac{\delta_{12}}{|\delta_{12}|} n_{12} \quad \text{and} \quad \delta'_{45} = \delta_{45} \pm n_{45}.$$

Under a change of coordinate system in $L_{4,4}$ the normal connections of the boundary a_{45} with the axis a_{12} must be carried to normal connections of a_{45} with the axis a_{23}. The result is a twist along the boundary a_{45} and the relation $\delta'_{45} = \delta_{45} \pm n_{45}$. We shall not go into the sign determination for this relation or the derivation of the various other coordinate transformations for the five-holed sphere. They are not important for our present purpose. However, the following is important : the coordinate system (a_{12}, a_{23}) in $L_{4,4}$ corresponds to the coordinate system (a_{45}, a_{34}) in $L_{4,1}$ under W_{14}, but the coordinate system (a_{23}, a_{12}) in $L_{4,4}$ corresponds to the coordinate system (a_{45}, a_{51}) in $L_{4,2}$ under W_{24}.

c) <u>Reduction of symbols.</u> By §7f) we can use a mapping of $L_{4,4}$ with movable boundaries to carry the curve system to one with the symbol pair

$$\begin{pmatrix} n'_{12} & n_{45} \\ \delta'_{12} & \delta'_{45} \end{pmatrix}, \text{ where } n'_{12} \text{ is smaller than } n_{45}.$$ In general the coordinate system remains unchanged in this reduction, it can remain as (a_{12}, a_{23}). However, when the curve system in $L_{4,4}$ consists of only two divisions, then it <u>may</u> be necessary to use the (a_{23}, a_{12})- coordinate system in place of (a_{12}, a_{23}) in order to achieve the reduction to $n'_{12} < n_{45}$. Finally, it is necessary to use the

(a_{13}, a_{12})-coordinate system in order to achieve $n'_{12} < n_{45}$ when $n_{12} = |\delta_{12}| > n_{45}$. We shall treat this exceptional case later. Then after reduction and the exchange W_{14} of four-holed spheres we have the symbol pair

$$\begin{pmatrix} n'_{12} \\ \delta'_{12} \end{pmatrix}\begin{pmatrix} n_{45} \\ \delta'_{45} & n'_{12} \end{pmatrix}$$

where n_{45} is now $> n'_{12}$ and where the three-termed symbol refers to the $L_{4,1}$ and the (a_{45}, a_{34})-coordinate system or, for the special case, to the $L_{4,2}$ and the (a_{45}, a_{51})-coordinate system. In each of these cases we can again reduce the symbol, so that $n'_{45} < n'_{12}$. We then take the corresponding exchange of four-holed spheres again, either W_{41} or W_{31} or W_{42} or W_{52} (the first exchange corresponds to the "general" case) and reduce in the new four-holed sphere. This process breaks off in case the number of boundary points in the three-termed symbol is equal to zero. We then have, say,

$$\begin{pmatrix} \bar{n}_{i,i+1} \\ \bar{\delta}_{i,i+1} \end{pmatrix}\begin{pmatrix} 0 \\ \bar{\delta}_{k,k+1} \end{pmatrix}.$$

This symbol denotes $|\delta_{k,k+1}|$ curves parallel to $a_{k,k+1}$ and $(\bar{n}_{i,i+1}, \bar{\delta}_{i,i+1})$ curves parallel to one of the three axes in $L_{k,k+1}$. Two of these three axes are of the form $a_{\ell,\ell+1}$ and are already determined. Still missing is the determination of the third, which is of the form $a_{\ell,\ell+2}$, where ℓ, $\ell+1$, $\ell+2$, k, $k+1$ are all different from each other modulo 5. We must determine the five axes $a_{\ell,\ell+2}$ in such a way that they do not intersect the axes $a_{k,k+1}$. This is done in Figure 35, in which the holes are reduced

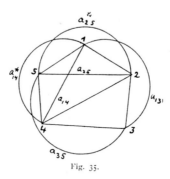

Fig. 35.

to points for the sake of clarity, the curves a_{ik} are given by doubly-traversed segments, and circuits around the holes are left out. The fact that a_{25} and a_{14} intersect at four points in our set-up is not important for these cases.

However, we still have to deal with the exceptional case, say $n_{12} = |\delta_{12}| > n_{45}$. Suppose that a_{13} is denoted by $\binom{1}{1}$ in $L_{4,4}$ with the (a_{12}, a_{23})-coordinate system, and that $n_{12} = \delta_{12}$, which we can always arrange by a twist along a_{12}. Then the symbol

$$\begin{pmatrix} n_{12} & & \\ & n_{45} & \\ n_{12} & & \end{pmatrix} \begin{pmatrix} n_{45} & \\ & \\ \delta_{45} \end{pmatrix}$$

represents curves which do not cut a_{13}. Among them, if $n_{12} > n_{45}$, there are $n_{12} - n_{45}$ closed curves parallel to a_{13}. The remainder are closed curves which are not parallel to boundaries of the L_4 cut along a_{13}, and hence capable of being mapped onto curves parallel to a_{24}, a_{54} or a_{25} by twisting along a_{24} and a_{54}. But the twist along a_{24} in $L_{4,5}$ can be composed from twists along a_{23} and a_{34} and the twist along the boundary a_{51}. In the exceptional case the fact that a_{52} and a_{14} intersect comes into play, in case our reduction process brings us to curve systems in $L_{4,4}$ or $L_{4,5}$. Because the axis a_{25} does not lie in the L_4 corresponding to a_{14} (in fact it intersects a_{14}), and a_{14} does not lie in the L_4 corresponding to a_{25}. Thus we choose two more curves, a_{25}^* which does not meet a_{14}, and a_{14}^* which does not meet a_{25} (see Fig. 35). We use these only in case we have to represent the curves in the L_4 corresponding to a_{14} or a_{25}. The exceptional case always appears only at the end of the reduction process.

A curve a_{ik} can be mapped onto no curve $a_{\ell m}$ for which one of the pair (i,k) equals one of the pair (ℓ,m). On the other hand, a mapping of the L_5 onto itself with movable boundaries which fixes two non-parallel, disjoint curves a_{ik} and $a_{\ell m}$ is a product of twists along a_{ik} and $a_{\ell m}$, because the latter curves divide the L_5 into three L_3's. Thus in summary we have:

1) Each mapping of the five-holed sphere with movable boundaries may be composed from twists along a_{12}, a_{23}, a_{34}, a_{45} and a_{51}, and by means of such a mapping each system of closed curves, apart from curves parallel to the boundaries, can be mapped onto closed curves parallel to two disjoint curves among the twelve curves a_{ik}, a_{14}^* and a_{25}^* (there are 15 such pairs).

2) The mapping class group for the L_5 is generated by linked even euclidean operations on the symbols $\binom{n}{\delta} m)$ and $\binom{m}{\delta} n)$.

§ 9.

Generation of the mapping class group for the L_n.[1]

We shall prove the following theorem:

Each mapping of an n-holed sphere belongs to the same class as a mapping composed from twists along fixed curves, to be chosen. The proof is derived from Lemmas 1) to 5):

1) Two closed curves which separate the same boundaries on the L_n are equivalent, i.e. they can be mapped onto each other by a mapping of the L_n with fixed or movable boundaries. Suppose, say,

---

[1] The mapping class group for the L_n, when the boundaries are points and can be exchanged, has been investigated by W. MAGNUS, Math. Ann. 109, p. 617 ff. Magnus finds not only a system of generators, but also a system of defining relations. Use is made of the connection between the mappings of L_n onto itself and the "braids" introduced by E. ARTIN. W. BURAU, Abh. Hamb. Math. Sem. 9, p. 117 ff. has given generators and relations for a subgroup of the braid group which corresponds to mappings of the punctured sphere with non-exchangeable boundaries. It may be conjectured that his results are related to ours.

that c_1 and c_2 both separate the boundaries $r_1, \ldots, r_m$ from the boundaries $r_{m+1}, \ldots, r_n$. Then by cutting the L_n along c_1 resp. c_2 we obtain the surfaces L_{m+1}^1 and L_{n-m+1}^1 resp. L_{m+1}^2 and L_{n-m+1}^2. By a mapping Φ, say, we can map the surface L_{m+1}^1 onto L_{m+1}^2 in such a way that the boundaries $r_1, \ldots, r_m$ on L_{m+1}^1 go onto $r_1, \ldots, r_m$ on L_{m+1}^2, and so that c_1 goes onto c_2. Moreover, we can map the surface L_{n-m+1}^1 onto L_{n-m+1}^2 by Ψ in such a way that $r_{m+1}, \ldots, r_n$ on the first surface go onto $r_{m+1}, \ldots, r_n$ on the second surface, and likewise so that c_1 goes onto c_2, in the same way that Φ maps c_1 pointwise onto c_2. Then Φ and Ψ together form a mapping (Φ, Ψ) of the whole L_n onto itself with non-exchangeable boundaries, carrying c_1 to c_2.

One proves similarly that: two self-connections of the same boundary which separate the same boundaries are equivalent. The same holds for two connections between the same pair of boundaries.

2) Two closed curves which separate the same boundary from all other boundaries are (not only equivalent, but also) homotopic. Because each of them is homotopic to the separated boundary, since each bounds jointly with it an L_2 on the L_n.

3) Two closed curves which are disjoint and which separate the same boundaries from each other are (not only equivalent, but also) homotopic. Because the two curves together bound an L_2 on the L_n.

4) When c_1 divides the boundaries into two groups, neither of which includes one of the groups enclosed by c_2, then c_1 and c_2 must intersect. In particular, we take c_1 to be a curve which separates just two boundaries, r_1 and r_2, from the others, and take c_2 to be an arbitrary curve which separates r_1 and r_2 from each other. By a homotopy of c_2 within c_1 or its neighbourhood, c_2 can be changed so that all parts of c_2 which run in the L_3 bounded by c_1, r_1 and r_2 become self-connections of c_1 which separate r_1 from r_2 (see Fig. 36). Since r_1 and r_2 now lie on different sides of c_2, the number of these self-connections

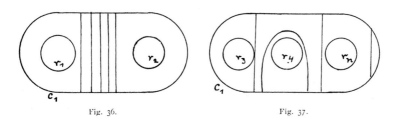

Fig. 36.   Fig. 37.

is odd and hence the number of intersections of c_1 with (the transformed) c_2 is not divisible by 4. Now we similarly consider the L_{n-1} (see Fig. 37) formed by the boundaries $r_3, \ldots, r_n$ on the other side of c_1 together with c_1 itself. Again we can use a homotopy to remove pieces of c_2 which are self-connections of c_1 not separating any of the boundaries r_i from the others on the L_{n-1}. Such a removal reduces the number of intersections of c_1 with c_2. The result is possibly more self-connections inside the L_3 on the other side of c_1 which do not separate r_1 from r_2, and which we can therefore remove. We repeat this process until each self-connection in the L_3 as well as in the L_{n-1} separates at least one boundary from the others. Thus we obtain, by homotopic transformation of c_2, a curve c_2^0 with $4m+2$ inter-sections with c_1 and $2m+1$ self-connections of c_1 in L_3 and L_{n-1}, of which each in L_3 separates r_1 from r_2 and each in L_{n-1} separates at least one boundary from the others.

We now consider another curve $\bar{c}_1$, disjoint from c_2, which again separates just two boundaries, say r_4 and r_5 from the others, but now suppose r_4 and r_5 <u>lie on the same side of</u> c_2. We apply the same homotopic transformation as produced c_2^0 from c_2 relative to L_3 to L_{n-1}, but now apply it to c_2^0 relative to $\bar{L}_3$ and $\bar{L}_{n-1}$, i.e. we remove all pieces of c_2^0 which are not boundary-separating self-connections in $\bar{L}_3$ and $\bar{L}_{n-1}$. Removal of self-connections in $\bar{L}_3$ does not alter the system of self-connections in L_3 and $\bar{L}_{n-1}$ at all. We shall not at first discuss the possibility that removal of self-connections in $\bar{L}_{n-1}$, i.e. replacement of parts of c_2^0 by segments which lie in $\bar{L}_3$ and therefore outside L_3 and disjoint from c_1, might result in the creation of self-connections in L_3 or L_{n-1}. If such changes did occur, then the number of intersections of c_2^0 with c_1 could only be <u>reduced</u>, and hence the number of self-connections in L_3 and L_{n-1} would be also. Then we could again apply the above process for L_3 and L_{n-1} and obtain a further reduction in the number of intersections between c_2 and c_1, until we again came to a curve, say c_2^{0*}, which had only boundary-separating self-connections in L_3 and L_{n-1}. Then we again operate with $\bar{L}_3$ and $\bar{L}_{n-1}$ and possibly reduce the number of intersections of the curve c_2^{0*} with $\bar{c}_1$. The process must terminate, and thus we finally obtain a curve c_2^{00} which gives only boundary-separating self-connections in $\bar{L}_3$ and $\bar{L}_{n-1}$ as well as in L_3 and L_{n-1}. Because of our hypothesis that r_1 and r_2 lie on different sides of c_2, while r_4 and r_5 lie on the same side, c_2^{00} <u>has numbers of intersections with both</u> c_1 <u>and</u> $\bar{c}_1$ <u>which are not divisible by</u> 4.

Now we shall convince ourselves of the fact, which was indeed not necessary for the construction of c_2^{00}, that application of the reduction process to c_2^0 relative to $\bar{L}_3$ and $\bar{L}_{n-1}$ cannot change the system of self-connections in L_3 and L_{n-1}. We have in fact the general theorem : <u>two simple closed curves on a sphere which cross form at least two 2-gons on each side of each curve.</u> The proof of this theorem follows immediately by consideration of the self-connections formed by one curve, say γ, in one of the two regions determined by the other curve; because the pairs of endpoints of such self-connections cannot separate each other on γ. Now a self-connection of c_2^0 in $\bar{L}_{n-1}$ which separates <u>no</u> boundary from the others and therefore must be removed, forms a curve γ' in combination with a piece of $\bar{c}_1$, dividing the sphere into two regions, one of which contains no boundary. If now each self-connection, and hence γ', were to intersect the curve c_1, then, by our general theorem, the self-connections of c_1 in this region would form with γ' at least two 2-gons containing no boundaries. However, a piece of γ' which forms a 2-gon with a piece of c_1 is a self-connection of c_2^0 in L_3 or L_{n-1} which separates no boundary from the others, and hence does not separate r_1 from r_2. Thus our process for the L_3 and L_{n-1} would not be terminated, contrary to hypothesis. Thus no self-connection of c_2^0 in $\bar{L}_{n-1}$ which is to be removed can meet the curve c_1, and hence the removal of such a self-connection of the system cannot alter the self-connections in L_3 and L_{n-1}.

5a) We consider a self-connection v of the boundary r and an arbitrary closed curve c. Then there will always be a self-connection v^0 between the same points of the boundary r, separating the same boundaries from each other, but <u>with two or no points in common with</u> c. Because c divides the L_n into an L_m and an L_{n-m+2}, and r lies on L_m, say. If v separates none of the boundaries lying on L_{n-m+2}, then there is a self-connection v^0 of r in L_m which separates the same boundaries on L_n as v and has no intersection with c. If on the other hand v separates boundaries on L_{n-m+2}, then there are first of all two connections of c and r, v_{11}^0 and v_{12}^0, which together separate the same boundaries on L_m as v. Likewise, on L_{n-m+2} there is a connection v_2^0, between the endpoints of v_{11}^0 and v_{12}^0 on c, which separates the same boundaries as v. In this case we obtain the desired self-connection v^0 of r as the polygonal path $v_{11}^0 v_2^0 v_{12}^0$ which twice cuts c.

b) Let v_1 and v_2 be two self-connections of the same boundary r whose endpoint pairs do not separate each other on r. Then it is proved as in a) that v_1 can be carried by a mapping of the L_n into a self-connection v_1^0 which has two or no points in common with v_2.

c) We consider a <u>system</u> $v_1, v_2, \ldots, v_m$, of <u>disjoint self-connections of</u> r, <u>and an arbitrary curve</u> c. In the part of r between the endpoints v_1 there can be no endpoints of $v_2, \ldots, v_m$, in the part of r between the endpoints of v_2 only the endpoints of v_1, etc. Some mapping of L_n takes v_1 to v_1^0 which has two or no points in common with c. This mapping takes $v_2, \ldots, v_m$ to $v_2^0, \ldots, v_m^0$. We construct a closed curve r^0 from v_1^0 and the part of r on which all the endpoints of $v_2^0, \ldots, v_m^0$ lie (see Fig. 38, in which r is mistakenly denoted by r_1 and v_1^0 by v_1).

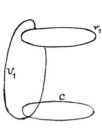

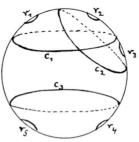

Fig. 38.   Fig. 39.

This closed curve bounds an L_m on the L_n which contains all self-connections $v_2^0, \ldots, v_m^0$. This L_m either contains no part of c, in which case $v_1^0, \ldots, v_m^0$ do not cut c at all, or else the whole curve c lies on L_m, or a part c^0 of c is a self-connection of r^0 with endpoints separating no endpoint pair of $v_2^0, \ldots, v_m^0$ on r^0. We now apply the considerations of a) resp. b) to c and v_2^0 resp. c^0 and v_2^0 and obtain, by a mapping of L_m with fixed boundary r_0, and hence also by a mapping of L_n, a mapping of v_2^0 onto v_2^{00}, where v_2^{00} has two or no points in common with c. In the process, v_1^0 is unaltered, $v_3^0, \ldots, v_m^0$ go to $v_3^{00}, \ldots, v_m^{00}$. We now handle v_3^{00} similarly and continue, finally obtaining the result : <u>the system</u> $v_1, v_2, \ldots, v_m$ <u>can be carried,</u> <u>by a mapping of</u> L_n, <u>onto a system of self-connections, each of which</u> <u>has at most two points in common with</u> c, <u>so that the system itself</u> <u>has at most</u> $2m$ <u>points in common with</u> c.

6) The theorem stated at the beginning has been proved for $n \leqslant 5$ (§§5,7,8). We shall now prove it for each n (first <u>without</u> determining the number of generators) with the help of complete induction. We therefore assume that it has been proved for L_m with $m < n$ and from this assumption prove it for $m = n$. We also assume that $n \geqslant 5$. We consider three curves c_1, c_2, c_3 (see Fig. 39) which separate the boundaries r_1, r_2 resp. r_2, r_3 resp. r_4, r_5 from the remaining boundaries. We choose these curves so that c_1 and c_2 have two points in common, and c_3 is disjoint from

c_1 and c_2. Moreover, we denote the L_{n-1}'s which c_1, c_2 and c_3 bound on the L_n by L_{n-1}^1 resp. L_{n-1}^2 resp. L_{n-1}^3. An arbitrary mapping Φ of L_n sends c_2 to γ_2; in general, γ_2 has points in common with c_3. We shall now construct a mapping Ψ^{-1} of L_n such that $\Psi^{-1}(\gamma_2)$ has no point in common with c_3, and therefore lies in L_{n-1}^3. Then, by 1), the curve $\Psi^{-1}(\gamma_2)$ can be carried to c_2 by a mapping Ψ_3^{-1} of L_{n-1}^3, so that $\Psi_3^{-1}\Psi^{-1}\Phi(c_2) = c_2$. I.e. $\Psi_3^{-1}\Psi^{-1}\Phi$ is a mapping (Ψ_2, Ω_2) composed from a mapping Ψ_2 of L_{n-1}^2 and a mapping Ω_2 of the three-holed sphere $(c_2, r_2, r_3) \equiv L_3^2$. Thus $\Phi = \Psi\Psi_3(\Psi_2, \Omega_2)$. By our induction hypothesis, Ψ_3, Ψ_2 and Ω_2 can be composed from twists along finitely many particular curves in L_{n-1}^3, L_{n-1}^2 and L_3^2. It remains only to compose the mapping Ψ^{-1} from twists along finitely many particular curves:

By 4), we can use homotopies to convert the curve γ_2 into the curve γ_2^{00} which forms no boundary-free 2-gons with c_1 resp. c_3. Then γ_2^{00} has $2m_1$ intersections with c_1 and $2m_3$ intersections with c_3, where m_1 is odd and m_3 is even. Thus $m_1 \neq m_3$; suppose $m_1 < m_3$. We then consider L_{n-1}^1, in which γ_2^{00} consists of m_1 self-connections of the boundary c_1. By 5), we can move these self-connections (with fixed boundary c_1) by a mapping Ψ_{11} of L_{n-1}^1 so that they have at most $2m_1$ points in common with c_3. Also, we can deform the curve $\Psi_{11}(\gamma_2^{00}) = \gamma_{21}^{00}$ homotopically (by 4) so that it forms no boundary-free 2-gons with c_1 and c_3.

In this way we obtain a curve γ_{21}^{00} which has $2m_{11}$ points in common with c_1 and $2m_{31}$ points in common with c_3, where $m_{11} \leqslant m_1$ and $m_{31} \leqslant m_1 < m_3$. Thus after possible homotopies, which we think of as combined into Ψ_{11}, the number of intersections with c_3 is reduced, the number with c_1 is not increased, and we still have $m_{11} \neq m_{31}$. If $m_{11} < m_{31}$ again, then we change γ_{21}^{00} by a mapping Ψ_{12} of L_{n-1}^1. However, if $m_{31} < m_{11}$, then we change γ_{21}^{00} by a mapping Ψ_{32} of L_{n-1}^3. In this way we obtain a curve γ_{22}^{00} which has $m_{12} \leqslant m_{11}$ points in common with c_1 and $m_{32} \leqslant m_{31}$ in common with c_3, <u>where strict inequality holds in at least one case</u>. We continue to so reduce the number of intersections with at least one of the two curves c_1 and c_3 until we finally have a curve γ_2^{000} with no more points in common with c_3. This attains the goal of finding the mappings Ψ_{ik} ($i = 1$ or 3) with which to compose the desired mapping Ψ^{-1} which makes γ_2 into a curve disjoint from c_3. By our induction hypothesis, the mappings Ψ_{ik} result from twists along particular curves in L_{n-1}^1 and L_{n-1}^3. Thus in summary we have: <u>each mapping of the L_n results by composition of mappings of</u> L_{n-1}^1, L_{n-1}^2 <u>and</u> L_{n-1}^3. (The mappings of L_3^2 onto itself can be regarded as mappings of L_{n-1}^3.)

7) As the first example of the general method we shall once again treat the L_5 (see Fig. 40).

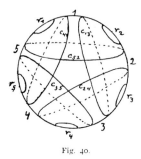

Fig. 40.

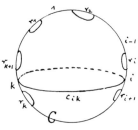

Fig. 41.

We denote the c_1, c_2, c_3 of the previous section more precisely by c_{52}, c_{13}, c_{35}, and correspondingly denote the L_4 which results from cutting along c_{ik} by L_4^{ik}. By the results on the L_4, the mappings of L_4^{35} (with fixed boundaries) are generated by twists along c_{52}, c_{13} ; c_{35}, r_1, r_2, r_3 ; the mappings of L_4^{52} (with fixed boundaries) by twists along c_{35}, c_{24} ; c_{52}, r_3, r_4, r_5 ; the mappings of L_4^{13} (with fixed boundaries) by twists along c_{41}, c_{35} ; c_{13}, r_4, r_5, r_1. Thus we have newly obtained the result that all mappings of the L_5 (with fixed boundaries) are generated by twists along c_{52}, c_{13}, c_{24}, c_{35}, c_{41} ; r_1, r_2, r_3, r_4, r_5.

8) In general we arrange the boundaries $r_1, \ldots, r_n$ of the L_n, as for L_4 and L_5, along a curve C which cuts each of them twice (see Fig. 41). The piece of C between r_i and r_{i+1} is denoted by C_i. We connect a piece C_i with a piece C_k ($i < k$) by a curve c_{ik} ; c_{ik} separates the boundaries $r_{k+1}, \ldots, r_n$, $r_1, \ldots, r_i$ from the boundaries $r_{i+1}, \ldots, r_k$. We choose the curves c_{ik} so that c_{ik} and $c_{i'k'}$ have two points in common if (i, k)

separates (i',k') , and no point in common if not. The corresponding curves on the $L_{(k-i)+1}$ and $L_{n-(k-i)+1}$ cut off by c_{ik}
are those already constructed for the L_n . Now since, as we have
already seen, all mappings of an L_5 , say, can be composed from
twists along curves of such a system, it follows from our construction of mappings of the L_n that the latter come from mappings
of three L_{n-1} 's and, by the induction step from $n-1$ to n , all
mappings of the L_n can therefore be generated by twists along the
$\frac{n(n-1)}{2}$ curves c_{ik} . Here the mappings are considered to have
fixed boundaries. If one allows the boundaries to move, then the
number of generating twists is reduced by the number of boundaries,
i.e. by n . We then obtain $\frac{n(n-3)}{2}$ generators. This result
holds for $n \geqslant 3$, while the result for the number of generators for
mappings with fixed boundaries holds for $n \geqslant 0$.

§ 10

The mapping class group for each orientable surface

1) By mapping an R_n^p onto itself, an arbitrary simple connection
v_{12} between two boundaries r_1 and r_2 can be carried to any
prescribed such connection v_{12}^0 .

2) For any two systems of disjoint simple self-connections v_1^i
and v_2^j of a boundary r_1 and a boundary r_2 on an R_p^n there is a
connection v_{12} between r_1 and r_2 which does not meet the v_1^i
and v_2^k : the self-connections v_1^i , in conjunction with r_1 , divide

the R_n^p into various simply or multiply connected regions.   However, none of these is bounded by lines v_1^i alone, because the latter are disjoint and not closed.   The boundary r_2 lies in one of these regions, say B.   The self-connections v_2^k of r_2 all lie in this region and, in conjunction with r_2, divide the region B into subregions.   The whole boundary of B belongs to one of these subregions, say $\bar{B}$.   Consequently, the boundary of $\bar{B}$ contains parts of both r_1 and r_2, which can therefore be connected by a curve not meeting the v_1^i and v_2^k.

3) We now suppose that we can already find generators for all R_m^q when $q < p$ (for $q = 0$ we have given this system in §9); and on the basis of this assumption we construct the system for R_n^p.   We divide the given R_n^p (see Fig. 42) by a curve C into an $R_{1,C}^1$ and

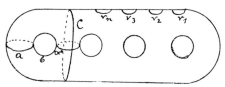

Fig. 42.

an $R_{n+1,C}^{p-1}$.   In the $R_{1,C}^1$ we consider a canonical cut system a, b. An arbitrary mapping Φ of the R_n^p onto itself carries a to α. We now construct a suitable mapping Ψ^{-1} such that $\Psi^{-1}\Phi(a) = a$.

Then $\Psi^{-1}\Phi$ is necessarily a mapping of the $R_{n+2,a}^{p-1}$ which results from cutting R_n^p along a, and hence a mapping whose generators are known by hypothesis. Consequently, our problem is to compose such a Ψ^{-1} from mappings of the $R_{1,C}^1$ and the $R_{n+2,a}^{p-1}$, which can be generated in turn by our induction hypothesis. It may be remarked that the mapping problem for an R_1^1 cannot be reduced in this way, because the C-curve always cuts off an R_1^1. However, the mappings of the R_1^1 have earlier been dealt with directly (see §b)).[1]

a) When the curve α crosses a, it is represented in $R_{n+2,a}^{p-1}$ by a system of self-connections of the boundaries a and $\bar{a}$ of $R_{n+2,a}^{p-1}$ and connections $v_{a\bar{a}}^i$ from one of these boundaries to the other. We choose one of the latter, say $v_{a\bar{a}}^1$. By 1), we can send it onto the connection b of a and $\bar{a}$ in $R_{n+2,a}^{p-1}$ by a mapping of $R_{n+2,a}^{p-1}$. Thus in this case there are (non-reducible) self-connections of the curve a in $R_{1,C}^1$ after the mapping, and hence by §6b) the number of intersections between α and a can be reduced by mapping the $R_{1,C}^1$. In this way we can reduce the number of intersections as long as the connections $v_{a\bar{a}}^1$ are present. Since the number of intersections between α and a is finite, this process must terminate, i.e. by alternate mappings of the

---

[1] For the once and twice punctured torus, R. FRICKE has already given generators and defining relations for the mapping class group. For this and the multiply punctured torus cf. W. MAGNUS, Math. Ann. 109. p. 634 ff. A system of generators for the mapping class group of the double torus R_0^2 has been given previously by the method of the arithmetic field, see M. DEHN, Autographie Breslau 1922 and R. BAER, J. f. Math. 160, p. 1 ff.

$R_{n+2,a}^{p-1}$ and $R_{1,C}^1$ we can remove all connections $v_{a\bar{a}}^i$. Suppose these mappings carry α to α^0.

b) If α^0 yields self-connections v_a^i of the boundary a and $v_{\bar{a}}^k$ of the boundary $\bar{a}$ on $R_{n+2,a}^{p-1}$, then by 2) we can connect a to $\bar{a}$ by a curve β which meets no v_a^i or $v_{\bar{a}}^k$. By 1), we then have a mapping of $R_{n+2,a}^{p-1}$ onto itself which sends β to b. Then the self-connections of a and $\bar{a}$ no longer have points in common with b. If we return from the $R_{n+2,a}^{p-1}$ to the R_n^p, then we have to join the boundaries a and $\bar{a}$ together. This possibly results in intersections of the self-connections with b, but by twisting along a we can arrange that this number is fewer than the number of intersections with a. By mapping $R_{1,C}^1$ onto itself, namely by exchange of a and b, we can then achieve a reduction in the intersections between α and a also. We continue in this way as long as the transformed α^0 yields connections between a and $\bar{a}$ or self-connections of a and $\bar{a}$. This process must terminate, i.e. we can reduce the number of intersections between α and a to zero. Suppose α^0 has now become α'.

c) Either α' is homotopic to a, and hence convertible to a by a homotopy, in which case we have reached our goal, or else it is non-homotopic to a, disjoint from a, and non-separating on R_n^p. On $R_{n+2,a}^{p-1}$ it is either non-separating or else it separates the boundaries a and $\bar{a}$, since it is non-separating on R_n^p. In both cases we use a mapping of $R^{p-1}_{n+2,a}$ to send α' to a curve α'' which cuts b once (see Figs. 42 and 43). A suitable neighbourhood

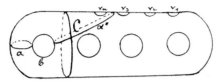

Fig. 43.

of the curve pair α'', b on the R^p_n constitutes an R^1_1. By twisting along α'' and b in this R^1_1 we can (§6b)) send α'' to b, and by twisting along b and a we can also send b to a (in $R^1_{1,C}$). But in each case the twist along α'' is a mapping of the $R^{p-1}_{n+2,a}$ onto itself. Thus in fact we have sent α to a by mappings in the $R^1_{1,C}$ and $R^{p-1}_{n+2,a}$, i.e. by a mapping ψ^{-1} with the desired property. We have now proved: <u>all mappings of a surface of genus p and n holes can be generated by twists along a finite number of particular closed curves</u>. In 5) we shall derive such a system of curves directly from the normal representation of the R^p_n.

4) <u>Examples</u>

a) R^2_0 (see Fig. 44). With our process in mind, we consider

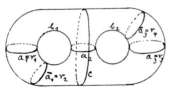

Fig. 44.

the R^1_{2,a_1} on R^2_0 and the $R^1_{1,C}$ containing a_1. Each mapping

of the R^2_0 onto itself is generated by mappings of these two surfaces

onto themselves. The mapping of the $R^1_{1,C}$ is generated by twists

along a_1 and b_1. In order to obtain the mappings of the R^1_{2,a_1},

we consider the $R^{1*}_{1,C}$ (with a_3 and b_2 as canonical cut system)

on R^1_{2,a_1} and the L_{4,a_1a_3}. All mappings of the R^2_{2,a_1} onto itself

are generated by mappings of these two surfaces onto themselves. The

mappings of $R^{1*}_{1,C}$ are generated by the twists along a_3 and b_2,

the mappings of L_{4,a_1a_3} are generated by twists along a_1 (and $\bar{a}_1$),

along a_3 (and $\bar{a}_3$), along a_2 and along C. However, (§6c), the twist

along C can be composed from twists along a_1 and b_1 (or also from

twists along a_3 and b_2). Thus we have: all mappings of the

closed double torus are generated by twists along the five curves

a_1, a_2, a_3, b_1 and b_2.

b) R^3_0 (see Fig. 45). Cutting the R^3_0 along a_1, a_4 and a_5

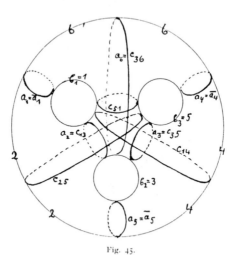

Fig. 45.

yields an L_6 whose boundaries may be denoted and ordered as follows

(see §9, 8) : $a_1 = r_1$, $\bar{a}_1 = r_2$, $a_5 = r_3$, $\bar{a}_5 = r_4$, $a_4 = r_4$, $\bar{a}_4 = r_6$.

We consider, as in §9, the generating twists c_{ik}, but leave out

c_{62}, c_{24}, c_{46}, because they can be generated by twists along a_1

and b_1, resp. a_5 and b_2, resp. a_4 and b_3. We have $c_{13} = a_2$,

$c_{35} = a_3$, c_{51} and also c_{14}, c_{25}, $c_{36} = a_6$. We claim that, on the

basis of the construction in 3), <u>all mappings of the triple torus</u>

<u>are composed from twists along the</u> 12 <u>curves</u> $a_1, \ldots, a_6$, b_1, b_2, b_3,

c_{51}, c_{14} <u>and</u> c_{25}. In fact, we consider the $R^1_{1,c_{62}}$ on the R^3_0

resulting from a cut along c_{62}, and the R^2_{2,a_1} resulting from a

cut along a_1. The mappings of $R^1_{1,c_{62}}$ are generated by twists

along a_1 and b_1. In order to obtain the mappings of R^2_{2,a_1}, we

consider the $R^1_{1,c_{46}}$ and $R^1_{4,a_1 a_4}$ on this surface. The mappings

of $R^1_{1,c_{46}}$ onto itself are generated by twists along a_4 and b_3.

In order to obtain the mappings of $R^1_{4,a_1 a_4}$ onto itself, we consider

the $R^1_{1,c_{24}}$ and $L_{6,a_1 a_4 a_5}$ on this surface. The mappings of

$R^1_{1,c_{24}}$ are generated by twists along a_5 and b_2, the mappings of

the L_6 are known by §9, 8 to be generated by twists along the

a_i, b_i, c_{51}, c_{14} and c_{25}. Thus our assertion is proved.

    5) In general, the closed surface R^p_0 is subject to the

following consideration : we choose p jointly non-separating curves

$a_1, \ldots, a_p$, and a second such system $b_1, \ldots, b_p$, where b_i has one

point in common with a_i, and none in common with a_k ($k \neq i$).

Cutting the R^p_0 along the a_i yields an L_{2p}.

By §9, 8, the mappings of this L_{2p} are generated by twists along $\frac{2p(2p-1)}{2}$ curves. But this counts the twists along the p curves a_i twice, since the twist along a_i is also the twist along $\bar{a}_i$. Moreover, there are another p twists, as we have previously remarked several times, which can be generated from twists along the a_i and b_i which on the other hand generate the mappings of the R^1_{1,a_ib_i} onto themselves. The curves $c_{2i,2i+2}$ (see below) along which these twists are made are all different for $p > 2$, whereas for $p = 2$ they all coincide. Thus the formula for the number of generating twists when $p > 2$ has to be reduced by 1 when $p = 2$. For $p > 2$ we have: <u>all mappings of the</u> R^p_0 <u>(which preserve orientation) are generated by twists along</u>

$$\frac{2p(2p-1)}{2} - 2p + p = 2p(p-1)$$

<u>curves.</u>

These curves consist, apart from the a_i and b_i, of curves which can easily be given when we arrange the boundaries a_i and $\bar{a}_i$ with the p curves b_i and another p connecting segments forming a closed curve, as above for $p = 3$. If we number these connecting segments successively by $v_1,\ldots,v_{2p}$, then the other curves for the generating twists are curves c_{ik} which meet each of two segments v_i and v_k at one point, where c_{ik} and $c_{i'k'}$ also have two resp. no points in common according as (i,k) and (i',k') separate each other or not.

Thus for the R_0^4 we obtain 24 generating twists in all.
Among these we have two twists along separating curves, namely
(see Fig. 46) along c_{48} and c_{26}. By §7f)2) the latter can be

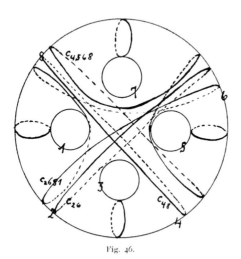

Fig. 46.

generated by twists along non-separating curves, which are given
in the figure as c_{4568} and c_{2681}.

6) We now add a short remark on the arithmetic field for the
L_n and R_n^p. One obtains the arithmetic field for the L_n by
dividing it into $n-2$ three-holed spheres $L_{3,j}$ by the $n-3$
closed curves c_j, so that the three-holed spheres are bounded by

$$r_1 r_2 c_1, \ r_3 c_1 c_2, \dots, r_{i+1} c_{i-1} c_i, \dots, \ r_{n-2} c_{n-4} c_{n-3}, \ r_n r_{n-1} c_{n-3}.$$

A curve system on the L_n with fixed boundaries is given by the intersection numbers and twist numbers for the n boundaries r_i and the $n-3$ curves c_j, and hence by $2n-3$ number pairs in all. With movable boundaries the n twist numbers for the boundaries drop out. If we divide the R_0^p by p jointly non-separating curves, then we obtain an L_{2p}. A curve system on the R_0^p yields a curve system on the L_{2p} for which the intersection numbers on any two corresponding boundaries are equal and for which the twist number along any one boundary in each pair can be taken equal to zero. Thus we obtain $3p-3$ number pairs in all from the above determination of curve systems on an R_0^p. Correspondingly, the curve system on an R_n^p is determined by $3p+2n-3$ number pairs.

In §5e) we have proved that the three number pairs which determine a curve system on the three-holed sphere are unchanged by homotopies. One concludes from this that the system of number pairs which we have just given for curve systems on the most general surface is also unchanged by homotopies.

APPENDIX : THE DEHN-NIELSEN THEOREM

JOHN STILLWELL

1. Introduction

One important theorem attributed to Dehn does not appear in his
published work, and his proof of it seems to be lost.   This is the
theorem that every automorphism of the fundamental group $\pi_1(S)$ of a
closed orientable surface S can be induced by a homeomorphism of S.
The earliest known proof appears in Nielsen [1927], and consists of two
parts.   The first, which is in §9 of Nielsen's paper and expounded
below as Theorem 5, is quite elegant and is attributed to Dehn.   The
second, in §23, is agonisingly long and is apparently Nielsen's own,
though he still credited the theorem to Dehn.   As was mentioned in the
introduction to Paper 7, Nielsen in 1931 found a replacement for §23
which was much simpler, but he did not publish it.   We give a similar
proof below as Theorem 6.

It is a pity that Nielsen did not publish his 1931 proof, as his
[1927] proof was made obsolete by the much shorter one of Seifert [1937],
with the result that Dehn's contribution to the theorem was buried along
with Nielsen's.   Moreover, since Seifert's proof was purely topological,
it became forgotten that the roots of the Dehn-Nielsen theorem were in
hyperbolic geometry.

With the recent revival of interest in hyperbolic geometry, it seems
timely to reconstruct the Dehn-Nielsen proof and its geometric background.
This not only fills a gap in the record of Dehn's work, it also brings
to light some little known but influential work of Poincaré.   It is
fairly well known that Poincaré discovered the connection between hyperbolic geometry and surface topology in his work on fuchsian groups in the
early 1880's (for an English translation of these papers, see Poincaré
[1985]).   In a less well known (though often cited) paper, Poincaré [1904]

took up the idea again for purely topological motives.   He obtained
results about simple curves on surfaces by a method which turned out to
be crucial to the Dehn-Nielsen proof - the use of geodesics as canonical
representatives for homotopy classes of curves.   This method is explained
in sections 3 and 4 below.

It was Poincaré [1904] which inspired Dehn to use hyperbolic geometry
in his solution of the word and conjugacy problems for surface groups,
Dehn [1912a] (Paper 4 in this volume).   In the 1920's, Dehn was still a
little under the influence of Poincaré, judging by the remarks at the
beginning of Dehn [1921], but he was gravitating towards purely topological
methods.   The torch of hyperbolic geometry in topology was picked up by
Nielsen in his [1927], [1929], [1932], a monumental series of works which
have only recently been assimilated, in the reworking and extension of
Nielsen's theory by Thurston.   In surveying the development of geometric
methods in topology, the Dehn-Nielsen proof seems an excellent vantage
point from which to look back to Poincaré and forward to Thurston.

It should be mentioned that the hyperbolic metric used in the Dehn-
Nielsen proof can be replaced by a metric on elements of the fundamental
group, the purely combinatorial "word metric" introduced by Dehn [1912a]
(see the Introduction to Paper 4).   A generalisation of Dehn's Theorem 5
is proved using the word metric by Floyd [1980], and it seems clear that
the concepts in the second part of the Dehn-Nielsen proof (Theorem 6)
could similarly be replaced by combinatorial ones.   What this probably
means, however, is not that hyperbolic geometry is irrelevant, but that
the most relevant features of hyperbolic geometry are the combinatorial
properties of its regular tesselations.

2. The special case of the torus

Among the closed orientable surfaces of genus ≥ 1, the torus T (genus 1) is distinguished by having a natural euclidean structure. This is one reason for a separate discussion of T. The other is that T enables us to review standard facts about canonical curves, universal covering and Cayley diagram in a context which is free of geometric difficulties. When we move on, to a surface S of genus > 1, it will then be possible to concentrate on the geometric difficulties which arise from the hyperbolic structure on S.

The torus T contains a pair of <u>canonical curves</u> a,b with a characteristic property: they are simple curves which meet at exactly one point, 0, where they cross. Cutting T along a,b yields the <u>canonical polygon</u> T_0 for T, which is topologically a square whose edges originate from a and b as indicated in Fig. 1.

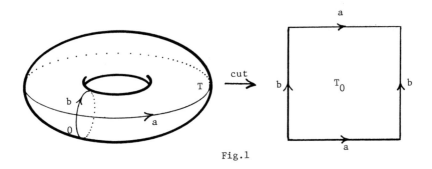

Fig.1

This follows from the classification theorem for surfaces (Dehn and Heegaard [1907]), and does not depend on the special position of a,b, as Fig.1 might suggest. Namely, the cut along b leaves a

connected orientable surface with two boundary curves (the two
"sides" of the cut, which are connected by the curve a). The
classification theorem says that any such surface is topologically a
cylinder, and cutting a cylinder along a simple arc a joining its
boundaries yields the square shown.

Thus if a', b' are any other curves with the characteristic
property of canonical curves, then cutting T along a', b' yields a
polygon T_0' homeomorphic to T_0, and with correspondingly labelled
sides. For example, Fig. 2.

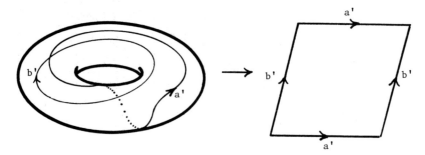

Fig. 2

It follows that <u>there is a homeomorphism</u> $\tau: T \to T$ <u>which maps</u> a   <u>onto</u>
a' <u>and</u> b <u>onto</u> b'. One only needs to construct a homeomorphism
of T_0 onto T_0' which maps equivalent points in the boundary of T_0
to equivalent points in the boundary of T_0'. A more intuitive way
to construct τ is to paste the equivalent sides of P' together
again to form a torus on which a', b' "look like" a, b (Fig. 3).

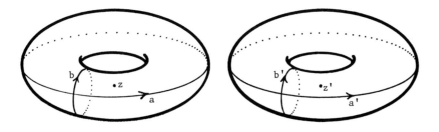

Fig 3.

Then the map τ: T → T which sends a, b onto a', b' respectively
is defined by sending each point z to the similarly positioned z'.

To summarise, we have

Theorem 1: If a, b are simple curves on T which meet at a single
point, where they cross, and if a', b' is another pair with the same
property, then there is a homeomorphism τ: T → T which maps a onto
a' and b onto b'.   □

It follows that if $\tau_*: \pi_1(T) \to \pi_1(T)$ is the automorphism of the
fundamental group induced by τ, then τ_* sends the homotopy class
[a] of a to the homotopy class [a'] of a', and the homotopy class
[b] of b to the homotopy class [b']. We are therefore
able to induce any automorphism I: [a] ↦ [a'], [b] ↦ [b'] of $\pi_1(T)$
for which [a'], [b'] are representable by a pair of canonical curves.
To prove the Dehn-Nielsen theorem for the torus, then, it suffices
that any automorphism preserves the characteristic property of canonical
curves.

Theorem 2. Any automorphism I: $\pi_1(T) \to \pi_1(T)$ can be induced by a
homeomorphism τ: T → T.

Proof: The fundamental group $\pi_1(T)$ is the free abelian group on two
generators (corresponding to a fixed pair of canonical curves a,b), so
we can identify it with the lattice of integer points (m,n) in the plane,

under vector addition.   An automorphism I: $\pi_1(T) \to \pi_1(T)$ is then a one-to-one linear map of the lattice onto itself, hence given by an integer matrix

$$I = \begin{pmatrix} p_1 & p_2 \\ q_1 & q_2 \end{pmatrix}$$

with determinant ± 1.   I sends the generators $(1,0)$ to (p_1, p_2) and $(0,1)$ to (q_1, q_2), where p_1, p_2 are coprime, and so are q_1, q_2, since det I = ± 1.   It can then be checked by brute force that the corresponding homotopy classes $[a^{p_1} b^{p_2}]$ and $[a^{q_1} b^{q_2}]$ can be represented by simple curves which meet, and cross, at a single point, whence the theorem follows from Theorem 1.   □

A much clearer view of the above proof can be obtained by looking at the universal cover of T.   I shall now sketch this approach, since it points the way to go in the case of higher genus, where the automorphisms of the fundamental group are no longer so accessible.

The universal cover $\tilde{T}$ of T is a plane obtained by pasting together copies of the canonical polygon T_0 for T so that the neighbourhood of each vertex looks like the neighbourhood of the single vertex 0 on T (Fig. 4).

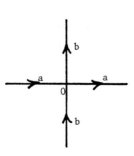

vertex neighbourhood

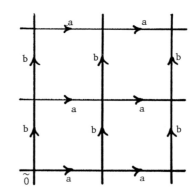

Fig.4

A point $\tilde{P}$ on $\tilde{T}$, relative to $\tilde{0}$, represents the homotopy class

[c] of an arc c on T with fixed endpoints 0 and P. $\tilde{P}$ is

found by lifting c to an arc $\tilde{c}$ on $\tilde{T}$ with initial point $\tilde{0}$, and

its association with the homotopy of class is due to the fact that a homotopy of

c with endpoints fixed lifts to a homotopy of $\tilde{c}$ with endpoints fixed.

Fig. 5 shows an example in which $\tilde{c}$ is deformed into the straight line

segment between $\tilde{0}$ and $\tilde{P}$.

T: 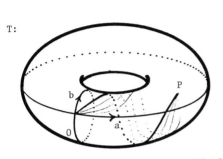 $\tilde{T}$: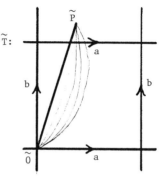

Fig.5

If we pull back the euclidean metric from the plane $\tilde{T}$ to the torus T,

then line segments on $\tilde{T}$ cover geodesics on T, and we can represent

each non-trivial homotopy class by a unique geodesic. If we imagine

that T is a self-contained world in which light rays travel along

geodesics, then $\tilde{T}$ is the <u>actual view</u> of T seen by its inhabitants.

In particular, the vertices, which we take to be situated at the

integer lattice points, are the different images of the origin seen

along closed geodesics. Thus the lattice points represent the homo-

topy classes of closed curves based at 0, the elements of $\pi_1(T)$,

with (m,n) corresponding to the element $[a^m b^n]$. This recovers the

representation of $\pi_1(T)$ as the integer lattice under vector

addition, and shows that the net of labelled edges on $\tilde{T}$ is the

Cayley diagram of $\pi_1(T)$.

An automorphism I: $\pi_1(T) \to \pi_1(T)$ can then be viewed as a one-to-one linear map $\tilde{f}$ of the lattice onto itself, and we can extend $\tilde{f}$ to a one-to-one linear map of the whole plane $\tilde{T}$. This extended map sends points (x,y) and $(x,y)+(m,n)$ over the same point P of T to points $\tilde{f}(x,y)$ and $\tilde{f}(x,y) + \tilde{f}(m,n)$ (by linearity of $\tilde{f}$) which again lie over the same point, $f(P)$ say, since $\tilde{f}(m,n)$ is a lattice point by hypothesis. Thus $\tilde{f}$ covers a homeomorphism $f\colon T \to T$, and f induces I because

$$[f(a^m b^n)] = \tilde{f}(m,n) = I[a^m b^n]$$

by definition of $\tilde{f}$.

Fig. 6 shows the $\tilde{f}$ and f associated with the automorphism I: $[a] \mapsto [ab]$, $[b] \mapsto [ab^2]$.

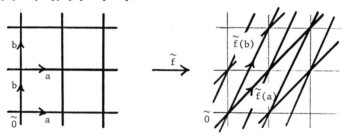

Fig.6

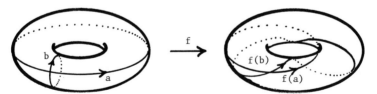

Although the above proof does not require an explicit determination of automorphisms of $\pi_1(T)$, it depends heavily on the ability of the euclidean plane to admit linear maps which are not rigid motions. This ability can be expressed algebraically by saying that any automorphism of the group of lattice translations extends to an automorphism of the group of all translations. In the hyperbolic plane there are no linear maps except rigid motions,[*] and hence there is no obvious way to extend the automorphism of the "lattice" which represents the fundamental group of a surface of genus > 1 to a map $\tilde{f}$ of the whole plane. In the absence of a map $\tilde{f}$ we have to make a rather exacting analysis of simple curves and intersections to show that an automorphism sends one canonical geodesic curve system to another.

[*] This can be seen from the projective model. In this model, the hyperbolic plane is the interior of the unit disc, "lines" are portions of ordinary straight lines within the unit disc, and rigid motions are all collineations of the ordinary plane which map the unit disc onto itself. Hence there are no non-rigid maps of the hyperbolic plane which map "lines" to "lines".

3. Facts from group theory and hyperbolic geometry

Hyperbolic geometry enters the theory of surfaces when one attempts to construct the universal cover $\tilde{S}$ of a closed surface S of genus $\geqslant 2$ in a metrically regular way. For example, cutting the surface of genus 2 along its canonical curves yields an octagon S_0 as canonical polygon (Fig. 7).

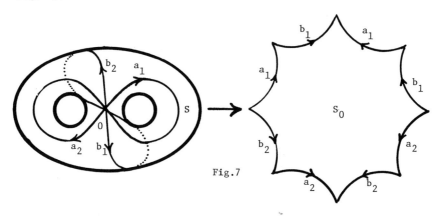

Fig.7

and $\tilde{S}$ is constructed by pasting copies of S_0 together so that each vertex has a neighbourhood like the neighbourhood of 0 on S, namely, Fig. 8

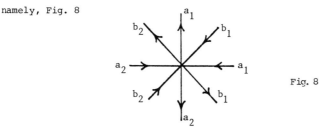

Fig. 8

Thus eight octagons meet at each vertex on $\tilde{S}$. This is not possible for regular euclidean octagons, but it is possible for regular octagons in the hyperbolic plane, where the angle sum of a polygon can take any value between 0 and the euclidean value. Thus one has regular octagons with corner angle $\frac{\pi}{4}$, and one can fit them together to form $\tilde{S}$ as a tessellation of the hyperbolic plane, the labelled edges of which form the Cayley diagram of $\pi_1(S)$. The construction is similar for any genus $p \geqslant 2$ - one has $4p$ $4p$-gons meeting at each vertex. Since $\pi_1(S)$ is the automorphism group of its own Cayley diagram, this gives a faithful representation of $\pi_1(S)$ as a group of hyperbolic motions. The motion associated with the element $g \in \pi_1(S)$ shifts the vertex $\underline{1}$ representing the identity of $\pi_1(S)$ to the vertex representing g, and hence the vertex representing h, for any $h \in \pi_1(S)$, to the vertex representing gh.

The topological properties of Cayley diagrams are said by Dehn 1922 to underly the proof of Dehn's theorem. To reconstruct such a proof I generalise the Cayley diagram construction to any curve system $\{\alpha_i\}$ which cuts S down to a simply connected piece. The curves α_i then lift to a net which partitions $\widetilde{S}$ into cells congruent to P. Since P is simply connected, each element of $\pi_1(S)$ is represented by a closed path through the union of the α_i on S, hence by a vertex of the net on $\widetilde{S}$ and by the corresponding motion of the net onto itself. To avoid making continual pedantic distinctions between elements $g \in \pi_1(S)$ and the vertices or motions which represent them, I shall speak of "the element g", "the vertex g" or "the motion g" as the occasion requires.

Of course, one can define $\widetilde{S}$ and its motions purely combinatorially, but the lesson to be learned from Poincare, Dehn and Nielsen is that hyperbolic metric concepts have great heuristic value, even if they can theoretically be eliminated. We shall therefore assume the basic facts of hyperbolic geometry, including the Poincare disc model, and develop only the less familiar facts which are pertinent to surface theory. The best sources for the facts we assume seem to be works on complex analysis, such as Siegel [1971], and Zieschang [1981].

Fact 1 (see Siegel [1971]).   The hyperbolic plane can be modelled by
the open unit disc $D \subseteq \mathbb{R}^2$ when "motions" are interpreted as conformal
(equivalently, circle-preserving) maps of $\mathbb{R}^2$ which map the unit circle,
∂D, onto itself.   Hyperbolic straight lines are then circular arcs
in D orthogonal to ∂D, hyperbolic angle equals euclidean angle,
and the motions are of three types corresponding to the following
families of euclidean circles in D (Fig. 9. (i), (ii), (iii))

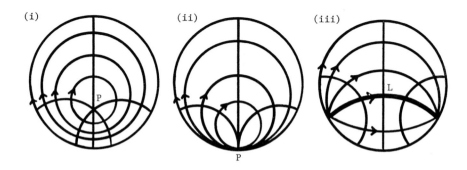

Fig.9

(i)   Circles orthogonal to the pencil of hyperbolic lines through a
point $P \in D$ are the hyperbolic circles with hyperbolic centre P.
A motion which leaves P fixed maps each circle with hyperbolic centre
P onto itself, permutes the hyperbolic lines through P, and is called
a rotation about P.

(ii)   Circles orthogonal to the pencil of hyperbolic parallels
with limit point $P \in \partial D$ are called hyperbolic limit circles with
centre P.   A motion which leaves P fixed maps each limit circle onto
itself, permutes the parallels with limit P, and is called a limit
rotation about P.   Any two hyperbolic parallels approach each other
arbitrarily closely in the hyperbolic metric, and under a limit
rotation there are points which are moved through arbitrarily small

hyperbolic distances.

(iii) Circles orthogonal to the hyperbolic perpendiculars to a given line L ⊆ D are the <u>curves at constant hyperbolic distance from</u> L (distance curves of L). A motion which fixes the end points of L on ∂D maps each distance curve of L onto itself, permutes the perpendiculars to L, and is called a <u>translation</u>, or <u>displacement</u>, with <u>axis</u> L. Each point of L is sent a constant hyperbolic distance, λ, called the <u>displacement length</u> of the motion, and the segment between any point of L and its image is called a <u>displacement segment</u>. Each point on a distance curve C of L is also sent a constant hyperbolic distance, but the constant is $> \lambda$ and increases with the distance of C from L.   □

<u>Fact 2.</u>   Each non-identity motion $g \in \pi_1(S)$ is a translation.

<u>Proof.</u> Since the motion g maps the Cayley diagram of $\pi_1(S)$ onto itself non-trivially, it must move every point through at least the diameter of a cell in the tessellation. Thus there is a non-zero lower bound to the distance moved by each point under the motion g, whence g is a translation by Fact 1.   □

As Dehn [1910] remarks, the non-existence of points on $\tilde{S}$ which move through arbitrarily small distances reflects the non-existence of small curves on S which do not contract to a point. This is why translations are important for topology, and we shall analyse them in detail. Following Nielsen [1944], we call the fixed points of a translation g its <u>fundamental points</u>; g moves points of D away

from the <u>negative</u> fundamental point, called U(g), towards the <u>positive</u>
fundamental point, called V(g), along the "streamlines" which are the
axis of g, called A(g), and its distance curves.

The next fact, also observed by Dehn [1910], gives a way to
find U(g), V(g), and hence A(g), from the Cayley diagram.

<u>Fact 3.</u>   The vertices ..., g^{-1}, 1, g, g^2,... of the Cayley diagram
lie on a distance curve through U(g), V(g).   Moreover $g^n \to V(g)$ and
$g^{-n} \to U(g)$ (in the euclidean metric), as $n \to \infty$.

<u>Proof.</u>   The motion g, by definition, maps each member of the
sequence of vertices ..., g^{-1}, 1, g, g^2,... onto its successor.   Hence
the vertices must lie along a streamline of the motion, i.e. along a
distance curve of A(g), by Fact 1.   Since they are equally spaced
along the distance curve, by Fact 1, the hyperbolic distances of g^n and
g^{-n} from 1 tend to ∞ as $n \to \infty$, hence $g^{-n} \to U(g)$ and $g^n \to V(g)$ in
the euclidean metric. $\square$

<u>Fact 4.</u>   If f, g are any two translations, then fgf^{-1} is a
translation of the same length as g, and

$$A(fgf^{-1}) = A(g) \cdot f^{-1}   \quad (\text{the } f^{-1}\text{-image of } A(g))$$

<u>Proof.</u>   Consider the f^{-1}-images, $U(g) \cdot f^{-1}$ and $V(g) \cdot f^{-1}$, of U(g)
and V(g), and their images in turn under fgf^{-1}.

$$U(g) \cdot f^{-1} \cdot fgf^{-1} = U(g) \cdot gf^{-1} = U(g) \cdot f^{-1}$$

since U(g) is fixed by g.   Thus $U(g) \cdot f^{-1}$ is a fixed point of
fgf^{-1}, and similarly so is $V(g) \cdot f^{-1}$.   Hence these are the fundamental

points of fgf^{-1} and $A(g) \cdot f^{-1} = A(fgf^{-1})$. Moreover, a point $P \in A(g)$ and its g-image $P \cdot g \in A(g)$ are sent by f^{-1} to $P \cdot f^{-1} \in A(fgf^{-1})$ and $P \cdot gf^{-1} =$ image of Pf^{-1} under fgf^{-1}, hence the translations on the two axes are of equal length. □

The remaining facts are proved following Nielsen [1927].

Fact 5. In a group of translations, two non-commuting elements have no common fundamental point.

Proof. Suppose on the contrary that f, g are non-commuting elements with a common fundamental point. By replacing one of f, g by its inverse, if necessary, we can assume that $V(f) = V(g)$. Now $V(fgf^{-1}) = V(g)$ by Fact 4, thus fgf^{-1} and g^{-1} have parallel axes (with limit $V(fgf^{-1}) = V(g) = U(g^{-1})$) and displacement lengths which are equal (by Fact 4) but opposite in direction. It then follows from the continuity of motions that by choosing a point P on $A(fgf^{-1})$ sufficiently close to $A(g^{-1})$, the point $P \cdot fgf^{-1} \cdot g^{-1}$ can be made arbitrarily close to P. Thus the non-identity group element $fgf^{-1}f^{-1}$ moves points through arbitrarily small distances, contrary to the properties of translations (Fact 1). □

Fact 6. In a group of translations, two elements commute if and only if they have the same axis.

Proof. If f, g have the same axis then $fg = [$translation of length $\lambda_f + \lambda_g$ with axis $A(f)] = gf$. Conversely, if $fg = gf$ then $fgf^{-1} = g$, hence $A(fgf^{-1}) = A(g)$. But $A(fgf^{-1}) = A(g) \cdot f^{-1}$ by Fact 4, hence f^{-1} must map the

hyperbolic line A(g) onto itself.   This line is therefore the
axis of f^{-1}, and hence of f, by Fact 1.   □

For the last fact we use Nielsen's notation I for an
automorphism and g_I for the I-image of g.

Fact 7.   An automorphism of a group G of translations induces
a permutation of the fundamental points of G.

Proof.   An automorphism sends $ghg^{-1}h^{-1}$ to 1 if and only if
$ghg^{-1}h^{-1} = 1$, hence $f_I g_I = g_I f_I$ if and only if   fg = gf.   Since
I permutes the elements of G, it then follows from Fact 6 that I
induces a permutation of the axes of G.   But no two distinct
axes have a common endpoint by Fact 5, hence the permutation of
axes is in fact a permutation of fundamental points.   □

4. Axes, simple curves and intersections

If we attempt to generalise the analysis of the torus given in section 2 to a surface S of higher genus, then several generalisations of the canonical geodesic curve system suggest themselves. For reasons which will appear later, it is good to take a system consisting of simple closed geodesics with no multiple intersections, and such that cuts along these geodesics render S simply connected. In fact, on the surface S of genus p there are geodesics α_1, α_2, ..., α_{2p} with the following properties

    (i)   Each α_i meets α_{i+1} at exactly one point, where they cross.

    (ii) There are no other multiple points; in particular, each α_i is simple.

We shall see in Theorem 3 below that cutting along such curves α_i renders S simply connected, so this system of curves meets all our requirements.

Such a system was used by Dehn's student Baer, [1927], and a similar one appears in Nielsen [1929, §4], though neither notes that the α_i can be realised by geodesics. Certainly it is clear that there are simple <u>curves</u> α_i with properties (i), (ii), e.g. as in Fig. 10, but to show that

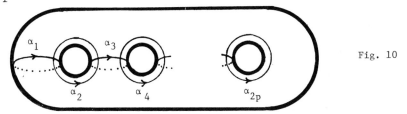

Fig. 10

topologically realisable incidences of curves are also realised by geodesics one needs the following ideas of Poincare [1904].

The metric on S is inherited from the hyperbolic metric on $\tilde{S}$, hence the geodesics on S are the curves covered by hyperbolic straight lines on $\tilde{S}$, and intersections of geodesics lift to intersections of hyperbolic lines. In turn, hyperbolic lines L_1, L_2 intersect if and only if the end points of L_1 on $\partial \tilde{S}$ separate the endpoints of L_2. Poincaré makes this trite remark topologically potent by combining it with the observation that a deformation of S lifts to a deformation of $\tilde{S}$ <u>through a bounded hyperbolic distance,</u> which therefore, <u>deforms any line into a curve with the same endpoints</u> on ∂S.

Lemmas 1-5 below systematise Poincaré's rather fragmentary applications of these remarks, and they suffice in particular to show the existence of geodesics α_i with properties (i), (ii). In stating the proofs it is convenient to extend the notation of section 3 to allow closed curves on S in place of their homotopy classes, the elements of $\pi_1(S)$. Thus when a closed curve c on S is lifted to $\tilde{c}$ on $\tilde{S}$ with initial point $\underline{1}$ we shall call the final point c. It is a vertex of the Cayley diagram, and the translation which carries $\underline{1}$ to c will be called the motion c, its axis $A(c)$, and the closed curve on S covered by a displacement segment $\delta(c)$ of $A(c)$ will be called $\alpha(c)$. (Of course, when we are discussing elements $g \in \pi_1(S)$ directly, the old notation will still be used, and $\alpha(g)$ will denote the curve on S covered by a displacement segment of $A(g)$.)

In what follows, "equivalent" sets on $\tilde{S}$ are sets with the same projection on S, i.e. translates of each other by motions in $\pi_1(S)$.

Lemma 1. The closed geodesics on S are precisely the curves $\alpha(g)$, $g \in \pi_1(S)$.

Proof. A closed curve c on S lifts to an arc $\tilde{c}$ on $\tilde{S}$ with endpoints equivalent under the motion c. If the curve c is a geodesic then $\tilde{c}$ is a hyperbolic line segment which is collinear with its own translate under the motion c. That is, $\tilde{c}$ is a displacement segment of A(c).

Conversely, a displacement segment $\delta(g)$ of A(g), $g \in \pi_1(S)$, has a closed projection $\alpha(g)$ on S, since the endpoints of $\delta(g)$ are equivalent (under g). And every segment of $\alpha(g)$ is geodesic because successive translates of $\delta(g)$ by g are collinear and equivalent (It follows that any displacement segment of A(g) has the same projection $\alpha(g)$, if we do not distinguish an initial point.) □

In general, the curve on S covered by a hyperbolic line segment with equivalent endpoints is a geodesic monogon with a corner where the two endpoints project, rather than a closed geodesic. For example, the curve covered by an edge labelled a_1 in the Cayley diagram of $\pi_1(S)$ has a right angled corner at the basepoint 0 on S, since adjacent edges labelled a_1 meet at right angles; thus segments of the curve which contain 0 are not geodesic. This monogon does however extend to a closed "figure eight" geodesic covered by adjacent edges a_1, a_2^{-1}, since such edges are collinear. In particular, the segment labelled $a_1 a_2^{-1}$ issuing from $\underline{1}$ on $\tilde{S}$ is a displacement segment of $A(a_1 a_2^{-1})$. (Cf. Figs. 7 and 8).

Since successive a_1 edges in the Cayley diagram are not
collinear, the distance curve of $A(a_1)$ through the vertices $\underline{1}, a_1, a_1^2, \ldots$
is not a hyperbolic line, and hence the curve a_1 on S cannot be
deformed into a geodesic with 0 fixed. This leads us to consider
free homotopies on S, i.e. deformations without fixed basepoint, and
their lifts to $\tilde{S}$ which are not fixed at the vertices of the Cayley
diagram.

Lemma 2 Any closed, non-contractible curve c on S is freely homotopic
to a unique closed geodesic, namely $\alpha(c)$.

Proof. Consider the lift $\tilde{c}$ of c which runs betwen the vertices $\underline{1}$ and
c in the Cayley diagram. Since the curve c is non-contractible on S,
these vertices are distinct and, since they lie on a distance curve of
$A(c)$, we can drop perpendiculars from them to the endpoints of a displacement segment $\delta(c)$ of $A(c)$. If we deform $\tilde{c}$ into the distance
curve segment between $\underline{1}$ and c, and then descend through the intervening
distance curve segments onto $\delta(c)$ (Fig.11), then the endpoints

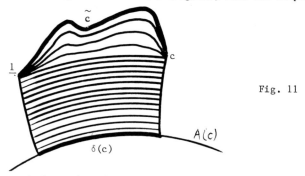

Fig. 11

remain equivalent throughout, and hence the whole deformation projects
to a free homotopy on S which converts c to $\alpha(c)$.

The uniqueness of $\alpha(c)$ can be seen as follows. By Lemma 1, any deformation of c to a closed geodesic lifts to a deformation $\tilde{c}_t$, $0 \leqslant t \leqslant 1$, from $\tilde{c}_0 = \tilde{c}$ to $\tilde{c}_1 =$ some $\delta(g)$, in which the endpoints of $\tilde{c}_t$ are equivalent for each t. Initially, when $\tilde{c}_t = \tilde{c}$, the endpoints are equivalent under the motion c, hence they must remain so (since $\pi_1(S)$ is a discontinuous group, e.g. by Fact 2). But the only axis which contains points equivalent under the motion c is $A(c)$, hence $\tilde{c}_1 = \delta(c)$, which projects to $\alpha(c)$ as required. $\square$

The next three lemmas exploit the separation properties of points on $\partial \tilde{S}$ and the fixture of these points under deformations on S.

Lemma 3. A curve c on S is freely homotopic to a simple curve $\Longleftrightarrow \alpha(c)$ is simple.

Proof. ($\Leftarrow$) is immediate from Lemma 2. To prove ($\Rightarrow$), suppose that $\alpha(c)$ is not simple. A self-intersection of $\alpha(c)$ is covered by an intersection of $A(c)$ with one of its translates $A(c) \cdot g$, $g \in \pi_1(S)$.

Case 1 : $A(c)$ crosses $A(c) \cdot g$. Then the endpoints of $A(c)$ separate those of $A(c) \cdot g$ on $\partial \tilde{S}$. We know from Lemma 2 that there is a deformation on S of $\alpha(c)$ to c, and this deformation lifts to a deformation on $\tilde{S}$ of $A(c)$ to a curve $\tilde{c}_\infty$ with the same endpoints on $\partial \tilde{S}$. The deformation similarly lifts to a deformation of $A(c) \cdot g$ to $\tilde{c}_\infty \cdot g$, with fixed endpoints. The endpoints therefore continue to separate each other on $\partial \tilde{S}$, and hence $\tilde{c}_\infty$ crosses $\tilde{c}_\infty \cdot g$ (e.g., by the Jordan curve theorem). This crossing covers a self intersection of c on S.

Case 2. $A(c) = A(c) \cdot g$, where g necessarily has axis $A(c)$ although it is not a power of c, since a double point on c lifts to points on $\tilde{S}$ which are inequivalent under the motion c. The corresponding deformations then yield $\tilde{c}_\infty$ and $\tilde{c}_\infty \cdot g$ with the same endpoints as $A(c)$. Since $\tilde{c}_\infty$ is periodic under the translation c along $A(c)$, it has points at maximum distance to left and right of $A(c)$, and since $\tilde{c}_\infty \cdot g$ is congruent to $\tilde{c}_\infty$ by the translation g along $A(c)$, $\tilde{c}_\infty \cdot g$ attains the same maximum distances to left and right of $A(c)$. It follows that $\tilde{c}_\infty \cdot g$ cannot lie entirely to one side of $\tilde{c}_\infty$, and hence they meet. Since g is not a power of c, an intersection of $\tilde{c}_\infty$ and $\tilde{c}_\infty \cdot g$ covers a double point of c. $\square$

Lemma 4. If c,d are disjoint simple curves on S, then $\alpha(c)$, $\alpha(d)$ are either disjoint or identical.

Proof. Axes, and hence geodesics, cannot be tangential unless they coincide, so if $\alpha(c)$, $\alpha(d)$ are neither disjoint nor identical they must cross, and hence $A(c)$ crosses $A(d)$ on $\tilde{S}$. Thus the endpoints of $A(c)$ on $\partial\tilde{S}$ separate those of $A(d)$. Deformations of $\alpha(c)$, $\alpha(d)$ to c,d respectively lift to deformations of $A(c)$, $A(d)$ to $\tilde{c}_\infty$, $\tilde{d}_\infty$, where $\tilde{c}_\infty$ has the same endpoints as $A(c)$ and $\tilde{d}_\infty$ has the same endpoints as $A(d)$. Then $\tilde{c}_\infty$ crosses $\tilde{d}_\infty$, and this crossing covers a point on S where c meets d, which is a contradiction. $\square$

The reader may easily guess that $\alpha(c) = \alpha(d) \iff c, d$ are freely homotopic, though we shall not need this result. It is worth pointing out, however, that the criterion $\alpha(c) = \alpha(d)$ then leads to an algorithm for deciding when two given curves are freely homotopic (Dehn [1910]), just as Lemma 3 leads to an algorithm for deciding whether c is homotopic to a simple curve (Poincaré [1904]).

Lemma 5. If c,d are simple curves which meet at a single point, where they cross, then $\alpha(c)$, $\alpha(d)$ have the same property.

Proof. We know from Lemma 3 that $\alpha(c)$, $\alpha(d)$ are simple.

To establish the crossing property, let $\tilde{c}_\infty$, $\tilde{d}_\infty$ denote the curves on $\tilde{S}$ which cover c,d and have the same endpoints as $A(c)$, $A(d)$. The single crossing X of c by d lifts to a single crossing $\tilde{X}$ of $\tilde{c}_\infty$ by $\tilde{d}_\infty$, together with a single crossing $\tilde{X} \cdot c^n$ of $\tilde{c}_\infty$ by each translate $\tilde{d}_\infty \cdot c^n$ of $\tilde{d}_\infty$ by a power of the motion c. The latter correspond to the encounters with X which result from successive traversals of c, and since c,d do not meet in any other way, $\tilde{c}_\infty$ does not meet any other translate $\tilde{d}_\infty \cdot g$, where $g \in \pi_1(S)$. We want to show that $A(c)$ is similarly crossed - by $A(d)$ and its translates $A(d) \cdot c^n$, and no other translates $A(d) \cdot g$ - as this will imply a single crossing of $\alpha(c)$ by $\alpha(d)$.

Since d is simple, all the translates $\tilde{d}_\infty \cdot c^n$ are disjoint, hence $\tilde{d}_\infty$ cannot lie along $A(c)$ by the argument of Lemma 3, case 2. Thus $A(d) \neq A(c)$ and the two axes must cross, otherwise there could not be a single crossing of $\tilde{c}_\infty$ by $\tilde{d}_\infty$. We therefore have a situation like that shown in Fig. 12.

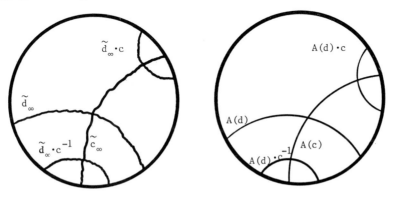

Fig. 12

Translates of A(d) by the motion c then also cross A(d), but translates by any other $g \in \pi_1(S)$ do not, since this would imply a crossing of $\tilde{c}_\infty$ by $\tilde{d}_\infty$ .g. □

It follows easily from Lemmas 3-5 that the curves $\alpha_1, \ldots, \alpha_{2p}$ on S with the properties (i), (ii) stated at the beginning of this section can be taken as geodesics. (The only worry is that disjoint curves might have coincident geodesic representatives, e.g. α_i and α_{i+2}. But this is ruled out because one of α_i, α_{i+2} crosses a curve which the other does not cross.) Any geodesics $\alpha_1, \ldots, \alpha_{2p}$ with properties (i), (ii) will be said to constitute a canonical geodesic system. The proofs of Lemmas 1-5 show that $\alpha_1, \ldots, \alpha_{2p}$ can be represented as $\alpha(g_1), \ldots, \alpha(g_{2p})$ for some $g_1, \ldots, g_{2p} \in \pi_1(S)$ whose axes have intersections which reflect the intersections of $\alpha_1, \ldots, \alpha_{2p}$. We combine these results into the following.

Theorem 3. For the surface S of genus p there are elements $g_1, \ldots, g_{2p}$ $\in \pi_1(S)$ such that $\alpha(g_1), \ldots, \alpha(g_{2p})$ is a canonical geodesic system on S. This means

(i) For each i, $A(g_i)$ crosses all translates $A(g_{i+1}) \cdot g_i^n$ of $A(g_{i+1})$ by the motion g_i, but no other translates $A(g_{i+1}) \cdot g$ where $g \in \pi_1(S)$

(ii) Except for the trivial coincidence of $A(g_i)$ with its own translates under the motions g_i^n, the different translates $A(g_i) \cdot g$, $1 \leq i \leq 2p$, $g \in \pi_1(S)$, have no other intersections. □

The above theorem does not appear explicitly in the literature, to my knowledge, but it would surely have been clear to Poincaré. Since Poincaré [1904] also proves the existence of homeomorphisms mapping curve systems with given intersections onto other systems with the same intersections, it also seems reasonable to attribute the following theorem to him.

Theorem 4. If $\alpha_1, \ldots, \alpha_{2p}$ and $\alpha_1', \ldots, \alpha_{2p}'$ are canonical geodesic systems on S, then there is a homeomorphism $\tau : S \to S$ such that $\tau(\alpha_i) = \alpha_i'$ for each i.

Proof. It is easily seen by induction on i that cutting S along $\alpha_1, \ldots, \alpha_{2i-1}$ gives a connected surface, with two boundaries that are joined by α_{2i}, and that cutting along α_{2i} gives a connected surface with one boundary. An Euler characteristic calculation then shows that when $i = p$ the cut surface becomes simply connected, i.e. a polygon P.

The boundary ∂P of P is divided into $8p-4$ segments identified in pairs : one pair for each of α_1, α_{2p} and two for each α_i, $2 \le i \le 2p - 1$, since each of the latter α_i is subdivided at two points, by α_{i-1} and α_{i+1}. We label and orient these segments by the names of the curves from which they originate, calling the pieces of α_i, $2 \le i \le 2p - 1$, α_{i1} and α_{i2}. The latter pieces are determined by the orientation of α_i if we let α_{i1} run from α_{i-1} to α_{i+1} and α_{i2} from α_{i+1} to α_{i-1} (See Fig. 14 below for the case $p = 2$.)

It can then also be checked by induction that the cyclic sequence of labels on ∂P is unique, hence the polygon P can be mapped onto the polygon P' which results from cutting S along the α_i' in such a way that each segment in ∂P is mapped onto the corresponding primed segment in $\partial P'$. Adjusting the map by a deformation if necessary, we can arrange that equivalent point pairs on ∂P are sent to equivalent point pairs on $\partial P'$, so that we have a homeomorphism $\tau : S \to S$ with $\tau(\alpha_i) = \alpha_i'$. □

Informally speaking, both P and P' can be pasted together to "look like" the standard picture of Fig. 10. This shows that there are homeomorphisms of S which send both the α_i and α_i' systems to standard position, and hence there is a homeomorphism of one system onto the other.

Let us now compare our situation to where we were with the torus after Theorem 1. Then it was clear, because the basepoint for $\pi_1(T)$ could be taken at the intersection of the two canonical curves a, b, that an automorphism I of $\pi_1(T)$ was "given" by the images a', b' of a, b, and it only remained to show that the I-images of a, b were canonical curves for any I.

Now, in avoiding multiple intersections, we seem to have lost a basepoint for $\pi_1(S)$, since there is no point common to all of $\alpha_1, \ldots, \alpha_{2p}$. However, since the α_i cut S down to a simply connected piece, it turns out that an automorphism I of $\pi_1(S)$ _is_ in fact "given" by its effect on the α_i. The main problem is showing that I sends one canonical geodesic system to another, when we know nothing about the automorphisms I.

This is where the avoidance of multiple intersections becomes important. We only have to show that I preserves the single intersection of α_i with α_{i+1}, for each i, and the non-intersection of α_i with α_j, $j \neq i - 1$, $i + 1$, which in turn reduces to showing that I preserves intersections and non-intersections of axes. Dehn proved this result very elegantly by looking at the behaviour of $\pi_1(S)$ at infinity. His idea is developed in the next section.

5. Mappings of $\partial\tilde{S}$ induced by automorphisms of $\pi_1(S)$.

As just mentioned above, the key step towards the Dehn-Nielsen theorem is to prove that an automorphism I of $\pi_1(S)$ preserves intersections and non-intersections of axes. We let I act on the set of axes by sending $A(g)$ to $A(g_I)$ for each $g \in \pi_1(S)$. Since $A(g)$ is determined by $U(g)$ and $V(g)$, it is equivalent to let I act on the set of fundamental points by sending $U(g)$ to $U(g_I)$ and $V(g)$ to $V(g_I)$. We know that this map is well-defined (in fact a bijection) by Fact 7, and it is appropriate to view the action of I on axes as really taking place on $\partial\tilde{S}$. In $\tilde{S}$ itself, I naturally acts on the vertices of the Cayley diagram by sending vertex h to vertex h_I, and this can only be interpreted as an action on axes by going to infinity (letting $h^{\pm n} \to U(h), V(h)$) as in Fact 3, we get $h_I^{\pm n} \to U(h_I), V(h_I)$).

It was Dehn's idea to investigate the action of I on axes by approximating axes by edge paths in the Cayley diagram. One then needs nothing but obvious properties of automorphisms in order to conclude that the map vertex $h \mapsto$ vertex h_I is a "quasi-isometry" of the Cayley diagram, and hence that nothing drastic can happen at $\partial\tilde{S}$, where the axes and their intersections are determined.

The approximation of axes by edge paths is implicit in Dehn's [1912a,b] solution of the conjugacy problem (see Nielsen [1927, §7(g)]), but Dehn's analysis of automorphisms seems to be lost, and the proof which follows is extracted from Nielsen [1927, §9], where credit is given to Dehn for the essential ideas.

Theorem 5 (Dehn). $U(g)$, $V(g)$ separate $U(h)$, $V(h)$ on $\partial \tilde{S}$
$\iff U(g_I)$, $V(g_I)$ separate $U(h_I)$, $V(h_I)$.

Proof. The first step of the proof is to approach $U(h)$, $V(h)$

by a doubly infinite edge path $p(h)$ through the vertices

h^n, $n = 0, \pm 1, \pm 2, \ldots$. We know from Fact 1 that these vertices

lie on a distance curve of $A(h)$. The segments of $p(h)$ between

them can be chosen with identical sequences of edge labels (spelling

out a word for h), and then $p(h)$ lies within a bounded hyper-

bolic distance of $A(h)$. We can similarly approach $U(g)$, $V(g)$

by polygonal paths through the vertex sequences $\{g^{-n} | n \geqslant N\}$ and

$\{g^n | n \geqslant N\}$, for any positive integer N. Assuming $U(g)$, $V(g)$,

$U(h)$, $V(h)$ are distinct, Fact 3 says we can choose N so that the

latter paths are arbitrarily far away from $p(g)$, and then if

$U(g)$, $V(g)$ separate $U(h)$, $V(h)$ we can also join their ends

g^{-N}, g^N by a path through successive vertices $g^{-N} = g_0, g_1, \ldots, g_i = g^N$

which are all at least as far from $p(h)$ as g^{-N}, g^N themselves.

The resulting path $p(g)$ is then far from $p(h)$.

We can also interpret "far" in terms of the word metric on the Cayley

diagram, which is the minimum number of edges connecting given vertices.

Vertices "far apart" in the hyperbolic metric are also "far apart" in the

word metric, since their hyperbolic distance is $\leqslant$ (number of edges × maximum

length of an edge), and conversely, since only finitely many vertices lie

within a given hyperbolic radius.

To simplify notation we denote the successive vertices of
$p(g)$ by $\ldots, g_{-1}, g_0, g_1, g_2, \ldots$ and the successive vertices
of $p(h)$ by $\ldots, h_{-1}, h_0, h_1, h_2, \ldots$. Fig. 13 shows how these
paths might look.

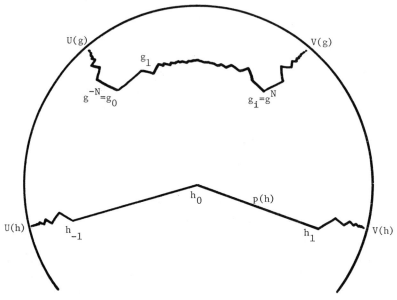

Fig.13

We now show that there are similar disjoint edge paths, $p(h_I)$
between $U(h_I)$ and $V(h_I)$, and $p(g_I)$ between $U(g_I)$ and $V(g_I)$, by
showing that I cannot send vertices which are far apart to vertices
which are close together. (This will show that $U(g_I)$, $V(g_I)$
separate $U(h_I)$, $V(h_I)$ if $U(g)$, $V(g)$ separate $U(h)$, $V(h)$, and
the converse follows by considering I^{-1}.)

In fact, letting $d(u,v) = $ length $(u^{-1}v)$ denote the minimum number
of edges in a path between the vertices u, v in the Cayley diagram
(the <u>word metric</u>) we shall show that there are constants k, $K > 0$
such that

$$d(u_I, v_I) \leqslant Kd(u,v) \quad \ldots \qquad (1)$$

$$d(u_I, v_I) \geqslant kd(u,v) \quad \ldots \qquad (2)$$

This says that I is a "quasi-isometry", while (2) in particular says that for any distance $d > 0$ in the word metric there is a $D > 0$ such that vertices $> D$ apart are sent to vertices $> d$ apart.

To prove (1), let w be a word of minimal length for $u^{-1}v$. Then

$$d(u,v) = \text{length } (w)$$

while $\quad d(u_I, v_I) = \text{length } (u_I^{-1} v_I)$

$$= \text{length } ((u^{-1}v)_I)$$

$$= \text{length } (w_I)$$

$$\leqslant K \text{ length } (w) = Kd(u,v)$$

where $K = \max \text{ length } \{(a_1)_I, \ldots, (b_p)_I\}$, since a word for w_I is obtained by replacing each generator a_i by $(a_i)_I$ and b_i by $(b_i)_I$. The inequality (2) now follows from

$$d(u,v) \leqslant K'd(u_I, v_I) \quad \ldots \quad (1)' ,$$

which is proved by applying the same argument to I^{-1}, and taking $k = 1/K'$.

Now to obtain a path $p(g_I)$ through the vertices $(g_i)_I$, and a disjoint path through the vertices $(h_i)_I$, we first notice, from (1), that each vertex $(g_i)_I$ can be connected to $(g_{i+1})_I$ by a path of length $\leqslant K$, since g_{i+1} is one edge away from g_i by definition. We construct $p(g_I)$, and similarly $p(h_I)$, as the union of such subpaths. Then to ensure that $p(g_I)$ does not meet $p(h_I)$ it is enough to ensure that each $(g_i)_I$ is distance $\geqslant 2K + 1$ from each $(h_j)_I$. By (2), this can be achieved if each g_i is distance $\geq D$ from each h_i, for a suitable D, and D in turn can be achieved by choosing a sufficiently large N in the construction of $p(g)$. $\quad\square$

The Dehn-Nielsen theorem now follows from Theorems 3, 4, 5.

Theorem 6. (Dehn-Nielsen) For any automorphism $I : \pi_1(S) \to \pi_1(S)$ there is a homeomorphism $\tau : S \to S$ which induces I.

Proof. Theorem 3 gives canonical geodesics $\alpha_i = \alpha(g_i)$ on S, and specifies intersections of the axes $A(g_i) \cdot g$, $g \in \pi_1(S)$, on $\tilde{S}$ which cover them. The axes $A((g_i)_I \cdot g_I$, $g_I \in \pi_1(S)$, which cover the geodesics $(\alpha_i)_I = \alpha((g_i)_I)$, are seen to have the same intersections, by Theorem 5, when one notes that they result from the $A(g_i) \cdot g$ by the I action.

Namely $\quad A(g_i) \cdot g = A(g^{-1} g_i\ g)$ by Fact 4, and
$$A((g^{-1} g_i\ g)_I) = A((g_I)^{-1} (g_i)_I\ g_I) = A((g_i)_I) \cdot g_I.$$

Thus the $(\alpha_i)_I$ also form a canonical geodesic system on S, and there is a homeomorphism $\tau : S \to S$ with $\tau(\alpha_i) = (\alpha_i)_I$ by Theorem 4.

In fact, the proof of Theorem 4 tells us that the polygons P, P_I which result from cutting S along the α_i and $(\alpha_i)_I$ systems respectively have isomorphically labelled boundaries, hence the generalised Cayley diagrams $\mathcal{D}, \mathcal{D}_I$ which result from lifting the α_i and $(\alpha_i)_I$ systems respectively to $\tilde{S}$ are isomorphic. Collinear edges of $\mathcal{D}$ unite to form the axes $A(g_i) \cdot g$, $g \in \pi_1(S)$, and collinear edges of $\mathcal{D}_I$ unite to form the axes of $A((g_i)_I) \cdot g_I$, $g_I \in \pi_1(S)$.

Fig. 14 shows the surface of genus 2 being cut along the α_i, and Fig.15 shows the cell of the Cayley diagram $\mathcal{D}$ which results.

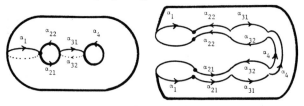

Fig. 14.

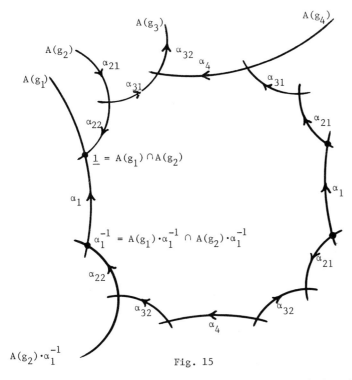

Fig. 15

If we let the identity vertex of $\mathcal{D}$ be $\underline{1} = A(g_1) \cap A(g_2)$ and let the identity vertex of $\mathcal{D}_I$ be $\underline{1}_I = A((g_1)_I) \cap A((g_2)_I)$, then the isomorphism $\mathcal{D} \to \mathcal{D}_I$ means that if an axis $A(h)$ is reached by following a certain sequence of α_i, α_{i1} or α_{i2} edges from $\underline{1}$ in $\mathcal{D}$, then $A(h_I)$ is reached by following the corresponding sequence of edges in $\mathcal{D}_I$ from $\underline{1}_I$. But the vertex of $\mathcal{D}$ representing $g \in \pi_1(S)$ is just

$$A(g_1) \cdot g \cap A(g_2) \cdot g = A(g^{-1}g_1g) \cap A(g^{-1}g_2g)$$

and hence g is determined by a common path to these two axes. The corresponding path from $\underline{1}_I$ in D_I leads to

$$A((g^{-1}g_1g)_I) \cap A((g^{-1}g_2g)_I) = A((g_I)^{-1}(g_1)_I g_I) \cap A((g_I)^{-1}(g_2)_I g_I)$$
$$= A((g_1)_I) \cdot g_I \cap A((g_2)_I) \cdot g_I$$

which is the vertex g_I of $\mathcal{D}_I$.

In other words, a closed path from $\alpha_1 \cap \alpha_2$ on S representing $g \in \pi_1(S)$ is mapped by τ onto a closed path from $\tau(\alpha_1 \cap \alpha_2) = (\alpha_1)_I \cap (\alpha_2)_I$ on $\tau(S)$ representing $g_I \in \pi_1(S)$. Thus τ induces the automorphism $I : \pi_1(S) \to \pi_1(S)$. $\quad \square$

References

Baer, R. [1927] : Kurventypen auf Flächen.

J. f. Reine und angew. Math. **156**, 231-246.

Dehn, M. [1910] : Papers 1 and 2, this volume.

[1912a]: Paper 4, this volume.

[1912b]: Paper 5, this volume.

[1922] : Paper 7, this volume.

Dehn, M. & Heegaard, P. [1907] : Analysis situs,

Encykl. Math. Wiss., vol. III AB3, Leipzig, 153-220.

Floyd, W. [1980] : Group completions and limit sets of kleinian groups.

Inv. Math. **57** (1980), 205-218.

Nielsen, J. [1927, 1929, 1932] : Untersuchungen zur Topologie der

geschlossenen zweiseitigen Flächen I, II, III. Acta

math. **50**, 189-358; **53**, 1-76; **58**, 87-167.

[1944] Surface transformation classes of algebraically finite

type. Det. Kgl. Danske Vidensk. Selskab, Math-fys.

Medd. XXI, 1-89.

Poincaré, H. [1904] Cinquième complément à l'analysis situs.

Rend. circ. mat. Palermo **18**, 45-110.

[1985] Papers on Fuchsian Functions. Springer-Verlag.

Seifert, H. [1937] Bemerkungen über stetigen Abbildungen von Flächen.

Abh. math. Sem. Univ. Hamburg **12**, 23-37.

Siegel, C.L. [1971] Topics in Complex Function Theory.

Vol. 2, Wiley, New York.

Zieschang, H. [1981] Finite Groups of Mapping Classes of Surfaces.

Springer-Verlag.